NEMESIS

From the Black Hole to the Infinite Universe
(with Donald Levy)

The Universe

Scientists Confront Velikovsky (ed.)

What is a Star?

The Search for Life in the Universe
(with Tobias Owen)

The Quest for Extraterrestrial Life (ed.)

The Evolving Universe

Cosmic Horizons
(with Robert Wagoner)

NEMESIS

The Death-Star
and
Other Theories of Mass
Extinction

Donald Goldsmith

Walker and Company
New York

First published in the United States of America in 1985 by
the Walker Publishing Company, Inc.

Published simultaneously in Canada by John Wiley &
Sons, Canada, Limited, Rexdale, Ontario.

Library of Congress Cataloging-in-Publication Data

Goldsmith, Donald.
    Nemesis: the death-star and other theories of mass
extinction.

    Bibliography: p.
    Includes index.
    1. Astronomy—Popular works. 2. Comets—Popular
works. 3. Extinction (Biology)—Popular works.
I. Title.
QB44.2.G65   1985   575   85-17785
ISBN 0-8027-0872-2

Printed in the United States of America

10 9 8 7 6 5 4 3 2 1

To Rachel,
who may yet find these
mysteries resolved.

CONTENTS

3: WHAT CAUSES MASS EXTINCTIONS?   83

NEMESIS

IN THE SPACE of one hundred and seventy-six years the Lower Mississippi has shortened itself two hundred and forty-two miles. That is an average of a trifle over one mile and a third per year. Therefore, any calm person, who is not blind or idiotic, can see that in the old Oölitic Silurian period, just a million years ago next November, the Lower Mississippi was upwards of one million three hundred thousand miles long, and stuck out over the Gulf of Mexico like a fishing rod. And by the same token any person can see that seven hundred and forty-two years from now the Lower Mississippi will be only a mile and three quarters long, and Cairo and New Orleans will have joined their sidewalks and be plodding comfortably along under a single mayor and a mutual board of aldermen. There is something fascinating about science. One gets such wholesale returns of conjecture out of such a trifling investment of fact.

—Mark Twain, *Life on the Mississippi*

PREFACE

THE "SHIVA" THEORY of life on Earth—that extraterrestrial objects have periodically bombarded our planet, causing mass extinctions every 25 to 30 million years—has gained prominence within the past few years, producing a wave of interest and a host of debates over the validity of different components of this multitiered hypothesis. Since the theory touches on important aspects of biology, geology, paleontology, dynamics, astrophysics, and the study of the origin and development of the solar system, these debates have involved a large number of scientists, each with something important to say about one or more parts of the theory. Anyone who seeks to write about the interplay of these scientific disciplines as exemplified in the discussion of the theories of mass extinctions has to work with a large and rapidly growing amount of data, as well as with an even larger amount of analysis of the data. Inevitably, the writer must rely on those who are willing to help search for the gem of truth that may or may not lie beneath the numerous theories.

In preparing this book, I have received generous assistance from those whom I asked for simpler explanations, detailed arguments and rebuttals, and material for illustrations. I would like to thank Luis Alvarez, Walter Alvarez, Frank Asaro, Barbara Bowman, Ken Brecher, Victor Clube, Stirling Colgate, David Cudaback, Marc Davis, Armand Delsemme, William Glen, Paul Goldsmith, George Gorman, Stephen Jay Gould,

Maurice Grolier, Jack Hills, Piet Hut, Erle Kauffman, Dennis Kent, Ivan King, Ed Krupp, Helen Michel, Alessandro Montanari, Nancy Morrison, Richard Muller, Carl Orth, Tobias Owen, Richard Pike, Michael Rampino, David Raup, J. John Sepkoski, Jr., Eugene Shoemaker, Michael Soule, Richard Stothers, Patrick Thaddeus, Richard Treffers, Scott Tremaine, and Paul Weissman for their kind assistance. My editor at Walker Books, Michael Sagalyn, convinced me that this book ought to be written, and has guided its preparation with skill and verve; I owe him a special debt of gratitude and wish him a long and successful career. My daughter Rachel played a key role in keeping me at work when I might have been otherwise occupied. To all of these people I can say: The help is yours, any errors mine. May we build on this foundation to a better understanding of how the world of science constructs hypotheses, plays with them, and selects the finest for the next generation.

INTRODUCTION

THIS BOOK TELLS the story of a scientific theory, developed during the past few years and now the subject of intense discussion in many parts of the scientific community. Like all theories in their youth, this one—that mass extinctions arise from periodic cometary bombardments—has evidence both for and against it; it has supporters and detractors; it is discussed by those who seek to make their names in science and by those who are already established, all of whom want to prove their ideas correct.

The theory contains several overlapping parts. I believe that some parts of the theory are almost certainly correct; others are quite speculative; and much of the theory falls in between, worthy of serious study but not at all proved to general satisfaction. To understand and to unsnarl the conflicting arguments has led me to a deeper enjoyment of the multifaceted cosmos in which we live—a joy that I hope to pass on to my readers.

In order to get a firm grasp on the different parts of the theory, I have divided it into four essential components. This book discusses each component in turn, saving the most controversial for last. My aim is to give the reader enough information to form his or her own opinion, based on scientific evidence. Each of the first three sections ends with a "rebuttal"—a presentation of some skeptical arguments against the sweeping assumptions of the theory. These rebuttals are not all-inclusive; other objections can be made, and have been made, to the different parts of the hypothesis that—if proved—will provide new insight into the history of life on Earth.

We may call this hypothesis the "Shiva theory," a name proposed independently by Harlan Smith of the University of Texas at Austin and by Stephen Jay Gould of Harvard University, after the Hindu god of destruction and rebirth. The theory consists of four linked hypotheses:

• First, that the great extinction which occurred 65 million years ago among many forms of life on Earth, including the dinosaurs, was the result of an asteroid or comet striking the earth.

• Second, that mass extinctions like these have recurred periodically, every 26 million to 33 million years.

• Third, that since there is no reasonable terrestrial cause of such periodic behavior, we must look outside the earth for an explanation of the cyclical extinctions; and that we find it in the trillions of comets orbiting the sun, each potentially capable of colliding with the earth.

• Fourth, that the most likely explanation of what drives the comets toward the earth's region of the solar system is either "Nemesis"—a companion star to the sun, named after the Greek goddess of divine justice—or else relatively close encounters of interstellar clouds of gas and dust with our solar system. The Nemesis star would be less massive and much less luminous than our sun and would move in an elongated orbit around the center of mass of the solar system, which lies close to our sun, by far the most massive solar-system object. Nemesis would perturb comets from their normally distant orbits by exerting its gravitational force on them. A close encounter of our solar system with an interstellar cloud would have a similar effect.

Trillions of iceberglike comets orbit the sun at distances tens of thousands of times the earth's distance from the sun. Gravitational perturbation of the orbits of these comets would thrust some of them into the inner solar system, and a tiny fraction of these interlopers could collide with Earth, producing one or more devastating impacts following each perturbation. The collisions would raise enormous clouds of dust, shrouding the earth in darkness for months on end and thereby extinguishing many forms of life. In a few years, the dust would settle out of the atmosphere. Life that had survived the cold and darkness would then renew itself and, freed from the competition of the vanished species, could spread and flourish, producing new species in a relatively brief time. In Hindu terms, Shiva, the god of destruction and renewal, would bring forth new life from the cometary catastrophe.

The Shiva theory draws on many scientific disciplines—in particular on geology, geophysics, paleontology, and astronomy. Hence one intriguing side effect of the theory has been communication among scientists from different disciplines who have never before engaged in debate with each other.

As we examine the Shiva theory, it is worth noting that not all of the theory's parts have the same weight of evidence or the same level of acceptance among scientists. The theory that a great impact led to the disappearance of the dinosaurs has gained widespread, though by no means total, acceptance. The hypothesis that mass extinctions recur in a periodic cycle has attracted many adherents, and seems capable of attracting more. The suggestion that such extinctions arise from cometary impacts stands on weaker ground, because calculations of the effects of such impacts remain uncertain, and there is no definite evidence that comet showers have occurred in past eras. Part four of the theory, which attempts to explain mass extinctions as the result of gravitational perturbations of the swarm of comets around the sun either by a solar companion star, Nemesis, or by interstellar clouds of gas and dust, is the most controversial of all. Disagreement exists not only between supporters and detractors of the Shiva theory in general but also between those who favor the two different theories of cometary perturbation.

If further research supports all four parts of the theory and shows what causes perturbations of comets that may strike the earth, then our view of the evolution of life on Earth—and, by implication, on other planets where life may exist—will change significantly. If mass extinctions occur on Earth every 26 million to 33 million years, they must have played a dominant role in biological evolution. The Shiva theory introduces a cyclical outside influence on biological evolution, on a time scale extremely long when compared with most evolutionary processes, but not longer than the existence of many successful species. The theory therefore creates an additional evolutionary pressure and evolutionary opportunity: If a species dies out during a period of cometary bombardment, it is gone forever, but if it survives, it finds tremendous new opportunity awaiting when the pall of dust has gone and the earth has again become "livable."

A second great potential consequence of the Shiva theory arises if the Nemesis hypothesis of a solar companion star proves correct. If the tremendous number of different species of life on Earth are the result of the existence of a low-mass solar companion, then evolution might proceed much differently, and far more slowly, in planetary systems lacking such an influence. We might have to envision one scenario for planets around single stars (which, until now, has always been taken to be the case for Earth and its fellow planets) and another for planets in a star system with a low-mass, comet-perturbing companion star. During the 13 billion years since the big bang began the expansion of the universe, the first scenario might have been able to produce much more highly evolved forms of life than the second, since only the first would include cycles of extinction followed by the rapid appearance of new species. Hence our speculation might be driven toward the conclusion that only in binary star systems, and specifically in those with one "normal" and one low-mass star, should we expect to find life as highly developed as life on Earth.

We shall save such speculations until we have examined the evidence for and against the Shiva theory. Let us now proceed to examine the record that nature has spread before us—first the evidence from the earth's crust and the fossil record held within, then from the starry skies around us. We shall find that whether or not the Shiva theory proves correct, we can learn from the debate a great deal that is relevant to the past, present, and future of life on Earth.

1

THE BIRTH OF THE IMPACT THEORY

EVERYONE KNOWS THAT the dinosaurs, lords of the earth for 150 million years, suddenly disappeared from the face of this planet. Throughout the Mesozoic era—which geologists subdivide into the Triassic, Jurassic, and Cretaceous periods—dinosaurs were the dominant land animals, ranging in size from primitive reptiles no larger than a chicken to the mighty animals whose names we learned as children: Diplodocus, Brachiosaurus, Ankylosaurus, Triceratops, Brontosaurus, and Tyrannosaurus rex. The fossil record shows that dinosaurs flourished for a period twenty times greater than that in which hominids have existed on Earth, and then became extinct. Within a geological instant, all the dinosaurs—hundreds of different species, and millions upon millions of individuals—vanished from the planet they had ruled for so long. Until the dinosaurs were gone, mammals spent tens of millions of years as shrewlike or opposumlike creatures, scurrying to survive amidst the domination of the earth by much larger dinosaurs. One of the most impressive films of my youth, Walt Disney's *Fantasia*, showed the gigantic dinosaurs dying in the desert as the earth's climate changed, leaving them helpless to confront an altered world. Did the dinosaurs die in this way, unable to adapt to a different environment? Did their sudden demise, which allowed mammals to radiate into a host of new species, occur within a year? A thousand years? A million?

THE EVOLUTION

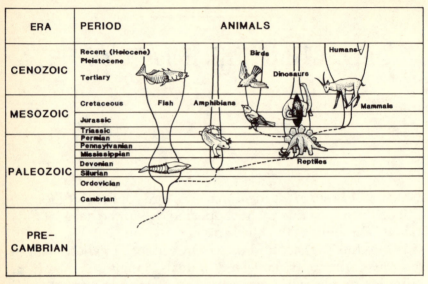

ERA	PERIOD	ANIMALS
CENOZOIC	Recent (Holocene) Pleistocene Tertiary	
MESOZOIC	Cretaceous Jurassic Triassic	
PALEOZOIC	Permian Pennsylvanian Mississippian Devonian Silurian Ordovician Cambrian	
PRE- CAMBRIAN		

Figure 1—The geological time scale for the evolution of life on Earth becomes complex after the vast diversification of living organisms at the end of the pre-Cambrian era, about 570 million years ago. Populations of land plants and animals grew significantly during the later Paleozoic era, between 400 and 300 million years ago. Mammals appeared during the Triassic period, at the beginning of the Mesozoic era, more than 200 million years ago, but did

The fossil record shows that thousands of dinosaur species evolved during the Mesozoic era, but it neither allows us to determine just *how rapidly* the dinosaurs became extinct nor permits us to state with certainty *why* they became extinct. All we know is that during the few million years that we can assign to the transition from the Cretaceous to the Tertiary geological periods, we pass from a time of large numbers of dinosaurs to a time of no dinosaurs at all. The cause of this disappearance—perhaps flooding from a vast, hypothetical Arctic lake; perhaps deadly radiation from a nearby exploding star; perhaps the drying up of the shallow seas as the result of climatic changes induced by a variety of possible mechanisms—remained a

OF LIFE ON EARTH

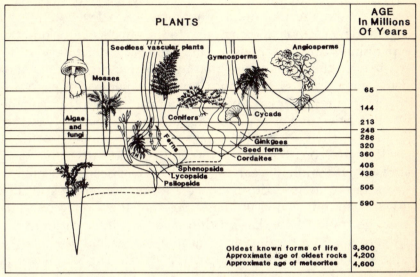

PLANTS	AGE In Millions Of Years

not diversify greatly until the dinosaurs vanished at the end of the Creta-ceous period, 65 million years ago, a time that also marks the transition from the Mesozoic to the Cenozoic era. The 150-million-year lifespan of the dinosaurs is represented by the shaded area within the section devoted to reptiles. (Drawing by Augusta Lucas-Andreae.)

subject for speculation, and little more than speculation, until the late 1970s, when geologists found a link in time betwen the extinction of the dinosaurs and the tremendous impact of a celestial object on Earth.

The Gubbio Layer

The Apennine Mountains form the spine of the Italian peninsula, not so high as the Alps, but significant mountains all the same, with a maximum elevation of nearly 3,000 meters at the Gran Sasso d'Italia. These mountains, like many others on Earth, consist of sedimentary rocks that formed many kilometers beneath the ocean surface, laid down over millions of years, layer by layer. A few million years ago, geological forces raised the mountain range above sea level, tilting its layers of rocks and exposing them to the effects of erosion. The Apennines consist largely of limestones: sedimentary rocks made from the shells of tiny marine animals, each a fraction of a millimeter across, that once housed single-celled sea creatures. In the Umbrian Apennines, near Gubbio, above the ancient cities of Perugia and Assisi, gather the streams that form the Tiber River. Here layers of limestones provide a geological record that spans the entire period from the Early Jurassic, 185 million years ago, to the Oligocene, 30 million years ago. And here, during the late 1970s, the geologist Walter Alvarez found evidence suggesting that the Cretaceous-Tertiary geological boundary, a thin layer of clay separating the Cretaceous from the Tertiary limestones, was laid down at the time of an extraterrestrial impact on Earth. This clay layer is now known as the Gubbio layer.

Protozoa That Build Rocks

Alvarez examined fossil foraminifera, the tiny creatures whose shells of calcium carbonate created the limestone rocks. These sea animals, typically smaller than a pinhead, live in all the earth's oceans, at all depths; they are amoebalike protozoa, the most common type of all sea life. Some "forams," as they are nicknamed, float freely in the oceans; others live in the bottom mud, spending their entire lives within an area of a few square centimeters. Their shells form chalk, limestone, and marble, and you can see foram fossils in the marble walls of public buildings if you examine them closely. The pyramids of Egypt are mostly limestone; they, too, consist of immense numbers of fossil foraminifera, so that a pharaoh's skeleton lay within trillions upon trillions of much older, more enduring

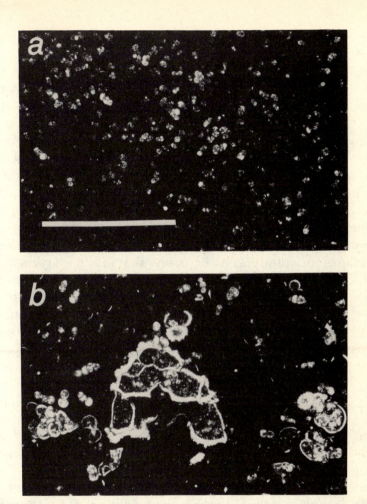

Figure 2—These photomicrographs, marked with a scale showing the length of a single millimeter, show fossil foraminifera from the rocks in the Bottaccione Gorge, near Gubbio Italy. The lower photograph (b), of rocks dating to the end of the Cretaceous period, shows a much wider variety and includes the large foraminifer *Globotruncana contusa;* the upper photograph (a), of rocks dating to the start of the Tertiary period of the Mesozoic era, shows only tiny foraminifera, *Globigerina eugubina*. (Courtesy of Profs. Walter Alvarez, Frank Asaro, and Helen Michel.)

skeletons of protozoa whose strength when linked together produces rocks that endure for millions of years.

Geologists have long been familiar with forams for their practical use: They serve as markers in looking for oil. A core drilled at a likely site can be sampled, dried, and studied for foraminifera. The different foram species are guides to the types of rock formations; hence the use of their name "oil bugs" among an older generation of petroleum geologists. Like other geologists, Walter Alvarez knew the importance of foram fossils when he cut his rock samples into thin sections and found fossils of different sizes, some only a small fraction of a millimeter, others nearly a millimeter in diameter. These were the planktonic (free-floating) rather than the benthic (bottom-dwelling) foraminifera. Alvarez saw that in those layers above the Cretaceous-Tertiary boundary—that is, in rocks formed less than 65 million years ago—none of the large forams appeared. He realized that he was looking at a foram extinction that must have occurred at the same time that the dinosaurs vanished—the demise of the millimeter-sized "giant" planktonic foraminifera.

Element Abundances in the Rocks of Gubbio

Alvarez took nine rock samples from the limestone layers in the Bottaccione Gorge, near Gubbio. Two of these samples came from layers immediately above and below the Cretaceous-Tertiary boundary; the other seven were from limestone rocks, ranging over 325 meters in height, in the Cretaceous-period rocks. In each of these samples, Alvarez and his coworkers, the nuclear chemists Frank Asaro and Helen Michel, proceeded to determine the abundances of twenty-eight different atomic elements.

The techniques used to determine element abundances owe much to the work of Alvarez's father, the Nobel prize–winning physicist Luis Alvarez of the University of California at Berkeley, where Walter Alvarez is also a professor. These techniques involve taking the rocks whose abundances are to be determined and bombarding them with neutrons accelerated to high velocities. Some of the neutrons strike atomic nuclei within these rocks and produce different types of radioactive nuclei. Each kind of radioactive nucleus "decays," falling apart into other types of nuclei in predictable ways. The new types of nuclei typically emit gamma rays (high-energy photons, the same type of particles that form light waves), and each new type emits gamma rays of a particular energy. By measuring the number of gamma rays of each energy that arise from the neutron bombardment, scientists can determine the number of each type of nucleus contained in even a tiny sample of matter.

Two of the world's experts at this technique are Frank Asaro and Helen Michel, who also work in Berkeley, at the University of California's Lawrence Berkeley Laboratory. Asaro and Michel have spent much effort analyzing the exact chemical composition of pottery fragments, to determine, among other things, trade routes of the ancient world. Because of the incredible accuracy with which element abundances can now be determined, Asaro and Michel can tell whether two potsherds came from the same geological stratum, the same general location, or even the same quarry. In one case Asaro and Michel found two fragments with such identical compositions that they could not resist speculating that the shards belonged to the same pot—whereupon someone gave it a try, and found that the fragments indeed fit together!

The Alvarezes persuaded Asaro and Michel to analyze their samples from the gorge near Gubbio. The chemists found that twenty-seven of the twenty-eight elements showed similar patterns of abundance variation from sample to sample. The twenty-eighth element was quite different. That element was iridium.

Iridium: Key to the Mystery

Iridium, element number 77, is extremely rare on Earth; it comprises only about one ten-billionth of the total material in the earth's crust. Iridium is one of the platinum group of elements, all of which bind easily with iron and are called siderophiles (meaning "iron-lovers"). Like its neighbor in the periodic table, osmium (element number 76), iridium has a particularly high density. Of all the elements, only iridium and osmium have densities more than 22 times that of water. Iridium and osmium are therefore even denser than platinum (21.4 times the density of water), gold (19.3), and uranium (18.7). Alvarez, Asaro, and Michel did not include osmium in their list of twenty-eight elements. They did include iridium, and measured its abundance in the samples from Gubbio. In the layer of clay a few centimeters thick that marks the boundary between the Cretaceous and Tertiary periods, they found that the fractional abundance of iridium rises to 41.6 parts per billion—that is, to an abundance 160 times greater than that found elsewhere in their samples.

From what geologists know about the history of Earth, and about the abundances of various elements in the sun and other stars, it appears almost certain that when our planet formed, it contained iridium in a much greater abundance than one ten-billionth, the typical fraction found in the crust today. This conclusion rests on evidence that the oldest meteorites—lumps of rock and metal left over from the formation of the solar system—whose composition is believed to resemble that of the primordial earth, have a fractional abundance of iridium about a thousand times larger than one ten-billionth. Because the early Earth was much more molten than it is today, the denser elements migrated toward its center. As a result, the Earth's core (the densest, innermost one-eighth of the earth's volume) consists largely of the denser elements such as iron and nickel.

The platinum-group elements also are apparently concentrated in the earth's core. In contrast, the mantle that surrounds the core, as well as the earth's outermost crust, consist mostly of the lighter elements silicon, oxygen, and aluminum. The core has an average density fifteen times the density of water, while the rocks in the crust and mantle average only three to four times the density of water.

What could produce the excess of iridium found at the Cretaceous-Tertiary boundary? The Alvarezes, Asaro, and Michel originally thought about nearby supernovas, exploding stars that spew vast quantities of material into space. But on closer examination this hypothesis fares poorly, since even a "nearby" supernova would be many light years from the solar system, so its output would be tremendously diluted by having to spread through a great volume by the time any of its material reached the sun and surrounding planets. Also, aside from the excess of iridium, the ratio of the different elements in the clay layer from Gubbio so closely resembles the element ratios in familiar solar-system material (in the earth, the moon, the sun, and meteorites) that it is highly unlikely such material could originate outside the solar system. Extra–solar system material would probably have different ratios of elements, which we could detect and recognize as too great to fit within the variations observed for solar-system objects.

Further thought by the Berkeley group suggested to them that the most likely way to bring iridium from outer space onto the earth would be to deposit the element directly onto the planet. Such direct transfer would involve a collision of cosmic dimension between Earth and an iridium-rich interloper.

The four scientists grew more convinced of the impact hypothesis once they traveled to Stevns Klint, a sea cliff about fifty kilometers south of Copenhagen in Denmark. There, too, layers of limestone (in this case chalk) include a record of the Cretaceous-Tertiary boundary. There, too, a layer of clay sev-

Figure 3—In the rocks of the Raton Basin of northern New Mexico and southern Colorado, the Cretaceous-Tertiary boundary clay appears as a thin dark layer, running through the center of this photograph from the lower left margin, under the ledge at left center, and on to the middle right, where it disappears beneath the vegetation. (Courtesy of Dr. Carl Orth.)

eral centimeters thick marks the boundary itself; and there, too, the abundance of iridium shows a marked rise in the boundary layer—in this case to a value close to 65 parts per billion. Once these results were in, the scientists embarked on a program to look for excess iridium worldwide, and they found it. By now more than seventy sites have been examined for iridium excesses, and such excesses have appeared at the Cretaceous-Tertiary boundary in all but three or four.

Many of the sites with excess iridium consist of sedimentary rocks formed underwater, but some have rocks that formed on land. For example, Carl Orth and his collaborators at the United States Geological Survey in Colorado have looked for iridium in the rocks of the Raton Basin of northern New

Mexico, which were not made on the sea floor, and they found a sharp rise—by a factor of 300—in a layer a few centimeters thick, at the level that marks the Cretaceous-Tertiary boundary investigated by the Alvarezes, Asaro, and Michel in the rocks from Gubbio and elsewhere. Still more intriguing, at just the same level Orth found a sharp *decline* in the number of fossil spores from angiosperms (flowering plants), a drop by a factor of 200 within a layer *just two millimeters wide.* This provides a close link in time between whatever event caused the iridium excess and a precipitous, temporary decline in at least one abundant form of plant life at the Cretaceous-Tertiary boundary, 65 million years ago. Hence the model of a catastrophe for many forms of life, caused by an iridium-rich object striking the earth, gains support from Orth's investigations. We are faced, as the Berkeley group was, with an astronomical question: What carried the iridium to Earth?

Iridium From Outer Space

No one suggests that a nugget of pure iridium vaporized in our atmosphere and deposited its matter as a shower of one particular heavy element. Such speculation violates what we know about the mixture of elements in the universe: A remarkable overall consistency, with only slight variations, appears from star to star and from place to place in the cosmos. The most reasonable proposal of how the iridium fell to Earth envisions an object that contained this "cosmic mixture" of elements, in which hydrogen, helium, oxygen, carbon, nitrogen, and silicon are the most abundant elements and iridium forms only a tiny part. (Rocky objects, such as the earth itself, have lost almost all of their hydrogen and helium but otherwise still reflect this mixture.) This tiny fraction, however, far exceeds the current iridium fraction in the Earth's *crust* because, as we have seen, the crust is relatively depleted in iridium as a result of the concentration of this element in the core while the earth was still largely molten. A cosmic mixture of elements—one like the mixture in the sun and in other stars whose element abundances are known from the elements' effects on emerging starlight—would contain nearly ten thousand times more iridium per gram than does the earth's crust.

We also know a good deal about the fractional abundance of iridium in extraterrestrial objects found on Earth: the meteorites, pieces of cosmic debris that have reached the earth's surface. Among meteorites, the oldest and least altered are the carbonaceous chondrites, believed to have a composition as close as we can find on Earth to the material that formed the solar system some 4.6 billion years ago. In the oldest of the carbonaceous chondrites (called Type I carbonaceous chondrites), the fractional abundance of iridium reaches 500 parts per billion, about two thousand times greater than the fractional abundance of iridium found in most terrestrial rock samples.

We can calculate how large an object striking the Earth and containing 500 parts per billion of iridium would have to be in order to leave behind the amount of iridium found in the Cretaceous-Tertiary boundary layer. This calculation should allow for the fact that not all of the material colliding with the earth would be distributed equally around it; some material would be deposited locally and would not show up, for instance, in the boundary layers of Gubbio and at Stevns Klint. Upon setting the fraction of material spread around the world at one-fifth, based on studies of the material ejected from the volcano Krakatoa's explosion in 1883, our calculation shows that in order to explain the excess of iridium at the Cretaceous-Tertiary boundary, we require an object with the composition of a Type I carbonaceous chondrite and a mass of about 350 trillion kilograms. This is the mass of a small mountain. To put things another way, if we imagine an object with a density of 2.2 grams per cubic centimeter, the same as the density of a carbonaceous chondrite, the object should have a diameter of

6.6 kilometers, the distance from City Hall to Rockefeller Center in New York City.

The Berkeley group therefore pictured an impact with the earth, 65 million years ago, of an object as old as the solar system and as large as lower Manhattan. This object would have deposited around the world the excess of iridium found in the Cretaceous-Tertiary boundary layer. It would also have presumably raised hell upon impact, extinguishing all the species of dinosaurs and many species of foraminifera, along with many other forms of life on this planet. Just *how* these extinctions would come about remains unclear, but no one can doubt that the collision of a 7-kilometer-wide object with the earth would be disastrous.

The Results of a Large Impact on Earth

Picture an object made of rock, clay, and dust, 5 to 10 kilometers across, colliding with Earth at a speed of about 15 kilometers per second. In less than a second, the object would sweep through the lower atmosphere, shoving the air aside to create an atmospheric "hole" along the object's path. An instant later, the object would strike either the ocean (a 70 percent probability) or the land. The difference in result would be less than might initially seem likely. Since the object will not be stopped until it encounters material with a total mass about equal to its own, the object could pass through 8 kilometers of water (the average depth of the oceans) while decelerating only partially. It would then bury itself in the rock below the ocean. If it struck on land, it would excavate a crater in the surface rocks of Earth.

Either impact event would create a huge crater, five to ten times the object's diameter, and would eject a vast quantity of matter upward and horizontally. In the case of an ocean impact, much of this matter would be sea water vaporized by the collision. For either kind of impact, most of the material ejected by the impact would be dust and grit, heated by the impact to hundreds or thousands of degrees, which would rise through the atmospheric hole just created. Much of the hot dust from the impact would diffuse around the stratosphere in less than an hour. Within a few hours after the collision, the earth would be surrounded by a heavy veil of atmospheric dust, and a long night would fall around our planet. The blanket of dust would slowly rain out of the upper atmosphere, restoring enough light for photosynthesis to resume after three or four months, revealing a world devoid of many of its former species, ripe for the development of new life forms in suddenly vacant ecological niches.

If this scenario sounds vaguely familiar, it should: We are discussing what is now generally referred to as "nuclear winter." The nuclear-winter scenario attempts to describe what would happen to the Earth after a large-scale nuclear war. Calculations made by the physicist Richard Turco, of R & D Associates in Marina del Rey, California; the atmospheric scientists O. Brian Toon, Thomas Ackerman, and James Pollack of the NASA Ames Research Center; and the astronomer Carl Sagan show that each nuclear explosion would put great quantities of soot and dust into the earth's atmosphere. These tiny particles would take months to settle back down, because they would remain above all the earth's weather; until they did, the earth would remain in near-total darkness.

Although extremely important in assessing the harmful results of a possible nuclear war, the nuclear-winter calculations deal with an effect *less* than that which could arise from an impact with a 7-kilometer object. Such an impact could make nuclear winter seem like Indian summer. We cannot now demonstrate that the dust from a giant impact would produce exactly those extinctions that occurred at the Cretaceous-Tertiary boundary, but the near-coincidence in time between the deposit of iridium and these extinctions makes it tempting to conclude that a casual link exists between the two sets of events and to assume as a working hypothesis that the iridium arose from an impact that also created the conditions that led to the extinctions. Some key questions immediately occur to the skeptics of new theories: From where did the presumed 7-kilometer object come? And where is the crater that it should have left behind?

Where Is the Impact Crater?

To answer the second question first, the proponents of the extinction-through-impact hypothesis would claim that the crater most likely lies beneath the oceans. Although 65 million years is sufficient time to erode a crater, it is probably not sufficient time for erosion to remove all traces of a crater as large as that made by an object 5 to 10 kilometers across, the sort of object hypothesized to have struck the earth with sufficient iridium to explain the excesses of that element found around the world.

A crater left by such an object would probably have a diameter greater than 40 kilometers. Giant, eroded craters of this size (and even larger) do appear on the earth's surface, and some are millions of years old. Indeed, these craters play an important role in any discussion of the hypothesis that periodic impacts cause periodic extinctions. But all the craters that we have found are on land; we do not yet have maps with the detail needed to detect relatively ill-defined crater basins far beneath the ocean. Futhermore, the collision of the earth's crustal plates (see pages 35–36), subducts some of their material, burying it as the plates move. This typically occurs below the oceans rather than on land. Indeed, some 20 percent of the Earth's crust (30 percent of the suboceanic crust) with an age of 65 million years or more has been subducted, depriving us of any chance ever to find the remnants of craters that the subducted crust might have carried.

Because we have an incomplete knowledge of the crust as it

was 65 million years ago, the lack of a place to point to and say, "There lie the remains of the Cretaceous-Tertiary crater," does not necessarily disqualify the impact theory. Luis Alvarez has said that when he began his work on the iridium excess, he expected that the point of impact could be located by determining where the iridium abundance was largest. However, the map of iridium excesses around the world shows no particular location of particularly high iridium concentration. Alvarez now ruefully states that he has learned that the dynamics of the explosion and the motions of the earth's crust provide a more complicated model than he had originally envisioned, but this in no way lessens his belief that a large object struck the earth some 65 million years ago.

Asteroids as Possible Impactors

The more demanding unanswered question from the impact theory concerns the *nature* of the object that struck Earth. When the Berkeley group first published their findings, in 1980, they were drawn toward the assumption that the impacting object would be an asteroid whose orbit crossed that of the earth. Since some asteroids are known to move in Earth-crossing orbits, this assumption seemed a straightforward application of the Occam's-razor principle dear to scientists: Do not multiply theories beyond necessity, for the hypothesis most likely to gain acceptance has the fewest invented components.

Asteroids—subplanets of the solar system—*do* exist in great numbers between the orbits of Mars and Jupiter; they range in size from Ceres, the largest, with a diameter of 1,000 kilometers, to the smallest one so far detected, a kilometer or so in diameter, and there is good reason to believe that many still smaller asteroids exist. Although most asteroids never pass inside the orbit of Mars, some *do* have smaller, more elongated orbits than the average asteroid, and some of these orbits cross the earth's. As a group, "earth-crossing" asteroids have been named Apollo asteroids. The best-known Apollo asteroid is Icarus, which has a diameter of about two kilometers and has approached Earth in this century to within two million kilometers (still five times the distance from the earth to the moon).

Like the other asteroids, Icarus is basically a huge rock, a piece of the solar system that never became part of a planet. For most asteroids—those orbiting between Mars and Jupiter—it appears that the perturbing effects of Jupiter's gravitational force kept the asteroidal "pieces" from ever coalescing to form a planet. These perturbations, as well as the lesser effect from the giant planet Saturn, may also be responsible for the unusual orbits of those asteroids that move in much more elongated trajectories than usual; some even intersect Saturn's orbit. The small-orbit asteroids with the most elongated orbits qualify as members of the Apollo group.

An asteroid five to ten kilometers across would qualify nicely as the object that caused the impact at the Cretaceous-Tertiary boundary. The chief difficulty with this hypothesis is that given the number of Apollo asteroids and the paths of their orbits, it would require a time span of hundreds of millions of years for such an impact to become reasonably likely. Therefore the hypothesis of such a collision as a random occurrence within the past 65 million years grows less reasonable on statistical grounds, though we certainly cannot eliminate it entirely.

We can now identify almost all the asteroids with diameters of seven kilometers or more; they number in the thousands, but not in the tens of thousands. Of these, the vast majority have orbits that stay comfortably between the orbits of Mars and Jupiter, so that the asteroids never come closer to the sun than 1.6 times the Earth's distance, and therefore remain much farther than Earth from the sun. Even if some of these asteroids approach still closer to the sun—closer than Mars—and even if some of these become Apollo asteroids, the chance of an encounter with Earth remains extremely low for any single asteroid, simply because the solar system is so vast in proportion to the sizes of the objects it contains. If we imagine the sun to be the size of a large light bulb at the center of a large living room, then the Earth would be a one-millimeter speck, orbiting the light bulb at a distance of 10 meters—well outside the room. Asteroids measuring one-thousandth of the earth's size would be invisible dust specks, highly unlikely to strike the earth even as they made millions of orbits around the sun.

This analysis does not eliminate the possibility of asteroids as impacting objects; it simply makes them improbable candidates for any given collision. If we look for a class of objects more likely than the asteroids to strike the earth, we can turn to objects so numerous that even though the chance of collision with any single one is nearly infinitesimal, the total probability of collision with one member of the class becomes significant— the comets.

Cometary Collisions With the Earth

On the morning of June 30, 1908, a tremendous fireball passed across central Siberia, creating an explosion that felled trees over an area of more than a thousand square kilometers, and a shock wave that went twice around the globe. The remoteness and inhospitality of the Tunguska area, in which the explosion occurred, prevented scientific investigation until 1930, when the Soviet geologist L. A. Kulik found vast areas of downed trees—all pointing away from the point of explosion— but no impact crater.

The exact nature of the "Tunguska event" remains uncertain, but the best guess is that a small comet almost struck the earth's surface that summer morning, vaporizing completely from friction in our atmosphere just before it could reach the ground. Even so, in less than a single second, the release of energy through this friction created the devastating localized effects that Kulik noted. In imagining an impact capable of causing mass extinctions, we must bear in mind that the Tunguska event released less than *one-millionth* of the energy that a 5-kilometer asteroid or comet would bring to Earth. This "tiny" amount of energy has led to speculation that the earth might have been struck by a piece of antimatter or by a "mini–black hole"; or, even more exotically, that an extraterrestrial spacecraft might have suffered a "nuclear drive catastrophe." These speculations aside, we can conclude that a comet *may* have hit the earth in 1908—but a small one in comparison to those that might cause mass extinctions.

Comets are the oldest pieces of the solar system, relics of the time when the sun and its planets formed, 4.6 billion years ago. Astronomers now believe that the larger objects in the solar system coalesced from bits of matter much like comets. These small bits first happened to collide and stick together; later, when they had formed objects with significant mass, these objects produced sufficient gravitational force to attract more bits with correspondingly increased force. The basic units in this process may be thought of as cometary nuclei, lumps of ice and dust a few kilometers, or a few tens of kilometers, in diameter. Most comets have spent the past 4.6 billion years in the celestial deep freeze, thousands of times farther away from the sun than is the earth. The number of comets thought to exist at these immense distances from the sun reaches beyond the thousands and millions into the *trillions*. Here we have a vast reservoir of objects that might collide with Earth. From spectroscopic studies of the way that comets reflect sunlight, we know that they have a composition something like that of the oldest known meteorites, the Type I carbonaceous chondrites, wrapped in a mantle of frozen ammonia, frozen methane, and frozen water.

There is one great problem in imagining that a comet could have collided with the earth at the time of the Cretaceous-Tertiary extinctions, and that is the enormous distances at which most comets orbit the sun. Only a few comets, diverted from their usual orbits by the gravitational forces from nearby stars, ever approach the realm of the planets and their satellites. The most famous such comet, named after Edmund Halley, has been diverted into an elongated orbit that stretches from inside the orbit of Venus to just outside the orbit of Neptune.

As Halley's Comet approaches the sun along this orbit, once every 76 years, the sun's warmth heats the comet, releasing some of the material in its nucleus. This matter diffuses around the comet, forming the fuzzy coma that surrounds the nucleus, and some of it streams behind the comet in its orbit, forming a gauzy tail that can stretch for millions of kilometers. In other words, Halley's comet is slowly wearing away, as each passage near the sun vaporizes a bit more of its material. Historical records spanning 2,500 years show that Halley's Comet has passed close to the sun several dozen times, without wearing away or breaking apart, and the comet must have made an unknown number of orbits before human recordkeeping began. Today the nucleus of Halley's Comet—all there is to it most of the time—has a diameter of about 10 kilometers. If it were to strike the earth, an event much like that hypothesized to have occurred 65 million years ago would take place, as an enormous cloud of gas and dust blanketed our home planet.

But what could make Halley's Comet strike the earth? And still more to the point, what could perturb large numbers of comets into Halley-like orbits, so that the probability of such collisions would increase to the point where such an impact became a real possibility rather than a tremendous long shot? The answer lies at the focus of the debate over the existence of the Nemesis star. In order to understand why the possible mechanisms for producing cometary impacts have such importance, we must step back and take a look at the broad sweep of the history of life on Earth. To do so requires that we examine the evidence that implies that the impact at the Cretaceous-

Tertiary boundary was not an isolated event in the history of our planet but, rather, one of a series of periodic, or quasi-periodic, times of impact that stretch over some 250 million years of the geological record.

Additional Iridium Anomalies

The Berkeley group has looked for other unusually high iridium abundances, which might be found at or near the geological layers that mark the times of other mass extinctions, such as those at the boundaries between the Paleozoic and Mesozoic eras (the Permian-Triassic boundary, about 248 million years ago) and between the Eocene and Oligocene periods, about 38 million years ago. In samples from sites in China and the United States, the Berkeley scientists found a thin layer of clay laid down at the Permian-Triassic boundary that is chemically quite distinct from the clay layers above and below it. They conclude that this clay probably formed from ash that rained down onto the seas and the land, but they did not find any excess iridium within the clay layer. The Berkeley group therefore suggests that the clay probably arose from volcanic eruptions rather than from a cloud of ash raised by a giant impact. Whether these eruptions might themselves have been caused by such an impact remains an open question; if an impact did occur, it seems to have left no iridium excess.

At the Eocene-Oligocene boundary, in a deep-sea drilling core taken from the bottom of the Caribbean Sea, Asaro and his colleagues did find an excess of iridium at just the position at which the extinctions of five species of radiolaria (microscopic sea creatures something like foraminifera, but with silicon-rich rather than calcium-rich shells) had previously been detected. Although the excess in iridium is not as pronounced as at the Cretaceous-Tertiary boundary, its presence again suggests that an extraterrestrial impact occurred, at a time when significant extinctions also took place in the oceans. Intriguingly, the iridium excess at the Eocene-Oligocene boundary has at least four distinct peaks in the iridium abundance, showing that four separate enrichments apparently occurred within a time of less than one million years.

In short, the evidence for other iridium-rich layers is suggestive but not so clear-cut as the enrichment at the Cretaceous-Tertiary boundary. Before we turn to the heart of the new theory—whether mass extinctions recur cyclically, and if so, what causes this recurrence—let us have a look at the arguments against the impact explanation of the extinction of the dinosaurs.

REBUTTAL: Did the Dinosaurs Die So Rapidly?

The Nature of Scientific Discussion

As we might expect—and indeed as is necessary for any scientific theory to make progress toward acceptance—the hypothesis that dinosaurs and other animals became extinct from the effects of an impact on Earth has drawn sharp criticism. Such criticism holds not that the hypothesis is impossible but, rather, that it is highly unlikely. Although some scientific hypotheses can be completely and utterly disproved to the satisfaction of all scientists, most cannot, since they deal with data too sparse and uncertain to be conclusive in a debate about a particular theory. This state of affairs, familiar to anyone who engages in scientific research, may contradict what nonscientists imagine, and hope, for the development of science, but in fact it is the only way scientists have found to make reliable progress: They debate the issues, lay them open for all to see, and rely on their analytical hypotheses to separate, in the long run, the more viable from the less viable theories.

Consider, for example, the debate over the theory of continental drift, the hypothesis that the separate pieces or plates of the earth's crust slowly move with respect to one another, and that as they grind against each other, with earthquakes more likely at their boundaries, they raise new mountain ranges. This theory was first proposed by Alfred Wegener, a German meteorologist. Because Wegener had no geological credentials, his theory drew little attention at first, and that attention consisted of scorn. But Wegener had an even greater difficulty than a lack of credentials; he had no mechanism to explain *why* crustal plates would move. Instead, he relied basically on the observation (not his alone) that the coastlines of the Americas fit nicely with those of Europe and Africa, as if the continents had been thrust apart at some time in the geological past.

Eventually, Wegener's hypothesis did gain confirmation, in two chief ways. First, geologists found the mechanism that explains why crustal plates move—sea-floor spreading. We now know that new material in the form of lava wells up from beneath the crust, forcing the adjoining plates apart. This spreading of the sea floor makes some of the plates ride over others to form mountains, causes other plates to move downward at the deep ocean trenches, and moves still others horizontally against each other, as the Pacific and North American Plates do at the San Andreas Fault. The spreading at the mid-Atlantic ridge has forced the Americas apart from Europe and Africa, just as Wegener imagined.

Second, and equally important for proving Wegener's hypothesis, detailed studies of the composition, age, and ancient magnetization of rocks from various locations have shown near-perfect matches between specimens now thousands of kilometers apart. This provides sufficient proof to all geologists that the rocks formed at the same place, and hence that *some* mechanism does separate continents over millions of years.

The continental-drift hypothesis thus provides a good example of how some theories make their way from seeming far-fetched but intriguing, as Wegener's did in 1915, to gaining near-total acceptance, as continental drift did by 1965. Some theories, of course, never gain wide acceptance; they are the ones scientists call "wrong." Others, bold though they appear at first, undergo sufficient testing to gain widespread acceptance within a single generation.

Einstein's special and general theories of relativity, proposed in 1905 and 1916 respectively, provide a good example of the rapid acceptance of a difficult, far-ranging set of hypotheses. By 1920, because the theories could be tested and passed that test, and because they provided a simpler framework to explain apparently disparate results, both theories had gathered a majority of adherents among prominent physicists. Within a few years after 1920, only some committed skeptics and a few anti-Semitic cranks continued to claim that Einstein was wrong. Einstein's theory of general relativity made an outstanding accomplishment that impressed many scientists: The theory not only accounted for many previously observed yet unexplained phenomena but also predicted a hitherto undetected phenomenon, the gravitational bending of light rays by the sun. When the British eclipse expedition of 1919 looked for and found such an effect, Einstein's name deservedly made the world headlines. Nothing can attract new adherents to a theory more stongly than suggesting that if we look for something, we shall find a particular set of events. Prediction carries much more impact than postdiction, the explanation of facts already known. Of course, without *some* postdiction, there can be no theories; scientists do not spend their time hypothesizing about completely unobserved phenomena. The trick is to build on what we think we know toward a new, more complete explanation—a new theory that will include what we have found so far and include a new hypothesis that adds to the realm of what we believe.

Most theories never get far off the ground; they stand as impressive tributes to human imagination but not as useful ways to interpret the universe around us. We should never forget, however, that when a scientist first puts forth a theory, he does not know into which class—useful or nonuseful, successful or unsuccessful, provable or disprovable—it will fall. As a result, for completely understandable reasons, a scientist who has a new hypothesis will push it as far as he or she can, first by internal, mental arguments, and then, if it appeals, in discussion with other scientists. Anyone who has ever engaged in such discussions knows that the proponent of a theory will push it as hard as possible; to do less would be to sell short one's own ideas. The give and take at, say, a scientific symposium involving contested theories therefore involves at least two simultaneous struggles: one for truth—that is, for a theory that will gain the widest acceptance—and the other for self-aggrandizement. This may sound a bit crass, but if you think about it, you may agree with me that we have here a wonderful example, through the so-called scientific method, of making individuals' desires to excel serve a more altruistic goal: the advancement of knowledge.

The Record of Dinosaur Fossils

The impact theory of extinctions at the Cretaceous-Tertiary boundary is a theory in its infancy, less than half a dozen years old. The wide-ranging nature of the hypothesis has, quite naturally, called forth a variety of criticism—not of its having been propounded (for all scientists love a bold new theory) but of its claimed validity. Foremost among the critics of the theory that an impact caused the extinction of the dinosaurs are the paleontologists J. David Archibald, of Yale University, and William Clemens, of the University of California at Berkeley (the home institution of Asaro, Michel, and the Alvarezes).

Clemens is what many of us once dreamed of becoming: an expert on dinosaur bones—how to find them, how to dig them out, how to put them together, and how to interpret their significance. Archibald is an expert on reconstructing the ecological situation of bygone eras. Both men have done extensive work in the Hell Creek Formation of eastern Montana, one of the great sites for finding dinosaur fossils. The first skeletons of Tyrannosaurus rex came from Hell Creek nearly a century ago; today, work there focuses on finding the fossils of other reptiles, mammals, amphibians, and plants. When these fossils were created, some 65 million years ago and more, eastern Montana lay at the edge of a sea that covered most of the land that the United States would later buy from France as the Louisiana Purchase.

From their detailed studies, Archibald and Clemens conclude that although it is clear that dinosaurs were gone from the earth by the time the Tertiary period began, they appear to have declined slowly all through the last 10 million years of the Cretaceous period, rather than vanishing all at once. In particular, Archibald and Clemens do not find dinosaur fossils reaching all the way up to the top of the layers representing the full Cretaceous period in Montana; instead, the uppermost dinosaur fossils tend to occur 2 to 3 meters below the layer that marks the Cretaceous-Tertiary boundary. These scientists do not, however, dispute Walter Alvarez's findings about the tiny foraminifera, which almost entirely vanish exactly at the Cretaceous-Tertiary boundary.

The Alvarezes, Asaro, and Michel have not been slow to consider and to respond to the argument made by Archibald and Clemens—that instead of seeing dinosaur fossils reaching right to the top of the Cretaceous layers and no farther, the geologists find the last dinosaur fossils in the Hell Creek beds 2 or 3 meters below the boundary with the Tertiary. The answer to this is simple: Since dinosaur fossils are in most cases fairly large, and since most dinosaurs have left no fossil remains, you cannot expect to find the last one you see in a vertical sample, such as that found in a rock face, precisely at the point where the dinosaurs became extinct. Consider, as an analogy, going north in the United States toward the Canadian border, and asking each person you met, "Are you Canadian or American?" In the normal course of events, the last "American" response would occur some modest distance south of the border, because people do not cover all of the landscape. Now consider the same question, but this time asked only of persons more than six feet four in height. These people represent the dinosaurs whose fossil remains have been discovered. Again, the last "American" response would occur to the south of the border, but in this case a good deal farther south than when you ask for *everybody's* nationality, since people over six feet four are much rarer than people in general. We can compare dinosaur fossils, which are found in modest quantities, to the people over six feet four in our analogy, and should expect to find the last such fossil somewhere below the Cretaceous-Tertiary boundary, rather than exactly at that point.

This analysis deals with the objection that the dinosaur fossils do not reach exactly up to the end of the Cretaceous, but it does not address the assertion by Archibald and Clemens that the density of fossils tends to thin out as one climbs toward the Cretaceous-Tertiary boundary, suggesting that dinosaurs were growing scarcer for many thousands of years before the time of extinction implied by the impact hypothesis. It may well be true that dinosaurs *were* growing less abundant at that time, for reasons unrelated to the approaching cometary impact.

Arguments From Probability

Archibald and Clemens suggest that dinosaurs vanished over a period of tens of thousands, or even hundreds of thousands of years, as demonstrated by their study of the fossil record at Hell Creek and elsewhere. If we take this as a working hypothesis, and refuse to connect the extinction of the dinosaurs with an impact on Earth 65 million years ago, it is extremely odd that the dinosaurs, by far the dominant land animals, all became extinct just before the time when the (proposed) impact occurred.

The dinosaurs lived for more than 100 million years. Even if they took as "long" as 100,000 years to become extinct, that would represent only one-thousandth of their total existence. Suppose we test the hypothesis that the impact and the dinosaur extinction are unrelated. If we picture the 100 million years before the hypothesized impact as a time chart 1 meter in length, we can throw a dart at the chart to represent the fact that the extinction does not depend on the impact. There would be only one chance in a thousand that the dart would land within 100,000 years of the time marking the impact. Hence, as Luis Alvarez would say, there is at most only one chance in a thousand that the dinosaur extinction is unrelated to the impact, if the latter indeed did occur, even before we ask just how the impact might have caused the dinosaurs to become extinct.

This argument cuts the other way, however, if we accept as true the statement by Archibald and Clemens that the dinosaurs were declining before their final extinction. Then only one chance in a thousand exists that their decline was unrelated to whatever finally made them extinct, because the time of decline would have occurred randomly within the 150-million-year span of the dinosaurs' existence. So unless we imagine "precursors" to a giant impact (perhaps mini-impacts from smaller objects), the statistical argument speaks against the impact hypothesis for the dinosaur extinction.

Arguments based on probability are rarely so satisfying, let alone so convincing, as those based on definitive evidence. It is therefore not surprising that Archibald and Clemens do not regard Alvarez's riposte as decisive. They prefer, quite understandably, to rely on what they can conclude from the dinosaur fossils, which they know as well as anyone does. In addition, they can cite studies of plant fossils by Leo Hickey, a paleobotanist at the Smithsonian Institution. Hickey's work shows that although some plant species did become extinct at the Cretaceous-Tertiary boundary, many other species did not, and that the time of such extinctions, instead of corresponding from place to place to within a time interval so short as to be unmeasurable on the geological record, varies by several tens of thousands of years from one location to another.

Thus some of the fossil experts, though ready to note that extinctions occurred within the last hundred thousand years of the Cretaceous period, tend to be dubious about any extinction so rapid as those suggested by the impact theory. These experts do not aim to contradict the hypothesis that a great impact occurred at the end of the Cretaceous. Indeed, Bruce Bohor and his coworkers at the United States Geological Survey have recently found additional evidence for such an impact: tiny quartz grains in the layer marking the Cretaceous-Tertiary boundary that show signs of "shock metamorphosis"—changes in structure caused by pressures much higher than those created even deep within the earth. Such quartz grains have been previously found only near impact craters from meteors and nuclear-bomb explosions. Not all geologists accept Bohor's interpretation of these grains, so once again we find a division of opinion among the experts. Further studies may soon uncover material that has definitely been altered by impact, thus supporting the impact hypothesis to the satisfaction of all—or they may not.

Let us take a quick look at a relatively recent extinction of several related species, an event for which we have detailed records because of its relatively recent occurrence: the extinctions of the giant mammals that we know and love from our trips to museums. These animal remains can be accurately dated by measuring the fraction they contain of carbon-14 nuclei, whose rate of decay is known. What do we find? The saber-toothed cat became extinct 14,000 years ago; the woolly mammoth 10,000 years ago; the giant ground sloth 8,500 years ago; the giant armadillo 7,800 years ago; and the colossal mastodon some 6,000 years ago. Another dozen mammalian species died out within this interval, which covers 8,000 years—a mere moment in geological history, but much longer than the theories based on nuclear-winter effect would predict. Of course, no one suggests that the giant mammals died because of a nuclear exchange, or because of a large impact, and their disappearance is not anything close to a mass extinction, but their gradual demise—perhaps because humans hunted them too far—does remind us that this extinction, the only one for which we have dates more accurate than a few thousand years, must arise from an effect quite different from those we have described. Evolution has always had a way of leaving a more complex record than sweeping theories would call for, and as we expand our knowledge about mass extinctions, we could well find that their explanations are far from identical. In examining the Shiva theory, we should bear in mind that a given set of events may not have a single explanation; nature may not be quite so simple as we would like it to be.

Did the Iridium Enrichment Arise From an Impact on Earth?

Two geologists from Dartmouth College, Charles Officer and Charles Drake, have a deeper doubt about the theory of extinction from a collision with an asteroid or comet: They dispute that the iridium-rich layer itself arose from any impact. Officer and Drake have studied the data on samples of the Cretaceous-Tertiary boundary layer taken both from land and from ocean-drilled cores (cylindrical sections thousands of meters long that are bored into the ocean floors and brought to the surface piece by piece).

After measuring the variation of the iridium abundance in the boundary layer, the geologists found that although in some locations the iridium peak shows a steep "spike," as if it were laid down instantaneously (in geological terms), at other locations the iridium peak is distributed over several tens of centimeters. Even though allowance must be made for the way that burrowing deep-sea animals (snails, for example) cause particles in a sedimentary layer to become somewhat vertically mixed, this effect amounts to no more than a few centimeters. Hence Officer and Drake conclude that the *tens* of centimeters of spread in the iridium-rich layers at some of the locations show the deposit of iridium over time periods ranging from 10,000 to 100,000 years. These intervals, short though they are in terms of the history of the earth, still far exceed the period of iridium-rich deposit that we would expect for iridium spread through the earth's atmosphere after a gigantic impact. Such iridium-rich particles should fall, and percolate to the ocean floors, within six months to a year, or perhaps a few years at the most, leaving the iridium-rich material concentrated within a layer only a few centimeters in depth.

If the strong enrichment of iridium did not result from an impact, what could have produced it? Officer and Drake favor a volcanic explanation. Although the theories of how the earth formed maintain that most iridium migrated to the earth's core, thousands of kilometers below the surface, it is possible that the abundance of iridium rises significantly below the crust, in the mantle, which forms the outer seven-eighths of the earth's volume (not counting the crust, a layer only a few kilometers thick that overlies the mantle).

Officer and Drake are properly impressed by a recent measurement of the iridium abundance in material ejected from the Kilauea volcano, on the island of Hawaii. When the geologists William Zoller, Josef Parrington, and Janet Phelan Kotra, of the University of Maryland, measured the element abundances in tiny particles collected in air filters during the volcano's eruption in January 1983, they found an abundance of iridium *ten to twenty thousand times* greater than normal. Therefore, even though no such iridium enrichment has been found in the matter from other active volcanoes (for example, the eruption of Mount St. Helens in 1980), Officer and Drake hypothesize that *some* volcanic eruptions tap iridium-rich matter, and that there is no need to look to the skies to explain a layer of iridium-rich material laid down 65 million years ago.

One intriguing speculation about the iridium abundance tends to unify the asteroid-impact and volcanic-eruption theories. Some geologists hypothesize that large impacts might trigger "hot spots" of volcanic activity. One of the most active hot spots, far from the plate boundaries where most volcanoes are located, has created the Hawaiian Islands, including the Kilauea volcano, during the past few million years. As the crustal movement dragged the Pacific plate slowly over the hot spot, each island appeared in succession; the main island, Hawaii, is the youngest of all. The older volcanic islands in this chain stretch through the Midway toward the Aleutian Islands, where the oldest of all lie buried in the Aleutian Trench. We therefore cannot determine just how or when the Hawaiian hot spot began its activity, and it remains possible, though completely speculative, that an impact 65 million years ago was the catalyst.

A skeptic would, however, point out that we have no need to invent impacts to explain volcanic hot spots, nor any need to look to impacts to explain the enhancement in the iridium abundance, if some volcanoes eject matter as enriched in iridium as the tiny particles recently taken from Kilauea. Dennis Kent, of Columbia University's Lamont-Doherty Geological Observatory, has studied the remains of Sumatra's Toba volcano, which erupted about 75,000 years ago. This volcano has a gigantic caldera (summit crater) that measures 35 by 100 kilometers, and apparently ejected four hundred times more material than the famous volcano Krakatoa, which exploded in 1883 in the same region of the world. Such an enormous volcanic eruption would throw an amount of material into the atmosphere comparable to that suggested for a 10-kilometer asteroid impact. If the volcanic matter came from the earth's mantle, and contained the same abundance of iridium as that in the hypothesized asteroid, then it could explain the iridium "spikes" found around the world as well as the impact theory can.

Officer and Drake propose that volcanic activity can explain all of the iridium enrichment, and that this is a much more likely source than an asteroid or comet. They cannot resist adding that although we have *no* evidence for an impact crater at the time of the Cretaceous-Tertiary transition, we *do* have evidence for an enormous vertical movement of matter from the Earth's mantle to its surface at that time. This evidence consists of the Deccan Traps of southwestern India, half a million square kilometers of flood basalts (igneous rock flows from volcanoes). The Deccan Trap volcanic activity lasted from 100 million to 30 million years ago but was especially intense near the time of the Cretaceous-Tertiary transition, 60 to 65 million years ago. Officer and Drake find that most of the rocks that form the Deccan Traps (now covered with topsoil and rich vegetation) were laid down in less than a million years. Clearly we ought to measure the iridium abundance there—a measurement not yet made—to see how sharp any sudden increase in the iridium abundance might be.

In view of the arguments made by Clemens and Archibald, by Hickey, and by Officer and Drake, we should not lose sight of the fact that these and other highly respectable geologists, paleontologists, and paleobotanists doubt that *any* extinction at the Cretaceous-Tertiary boundary took less than a few thousand years and that the iridium enhancement proves that an impact occurred at that time. The hypothesis that an extraterrestrial object hit the Earth 65 million years ago and caused worldwide, catastrophic extinctions remains a hypothesis, not a demonstrated fact. But given the iridium data, and the correspondence in time between the iridium enrichment and the Cretaceous-Tertiary mass extinction, it seems worthwhile to examine the impact theory further, if only to see whether it can be better tested. To do so, we embark on a more global inquiry: Was the impact at the Cretaceous-Tertiary boundary the sole event of its kind in the history of Earth? If not, can we assign a single cause—extraterrestrial impacts—to all such great extinctions? And do these impacts occur at random, or in a periodic cycle?

2

Do Extinctions and Impacts Come in Cycles?

THE EXTINCTION OF the dinosaurs represents just one, though by far the most widely publicized, of the numerous mass extinctions of life on Earth. Paleontologists have long known of several such episodes, the most drastic of which occurred 248 million years ago, at the end of the Permian era. This "Permian-Triassic extinction" eliminated *more than 95 percent* of the species then in existence; since most of these were marine invertebrates similar to sponges, jellyfish, and sea snails, their extinction has not caught the public's attention to the same extent as the demise of the dinosaurs at the end of the Cretaceous period. These mass extinctions, however, represent episodes in biological history quite different from ordinary extinctions. Some species go extinct during any long time interval, but periods of mass extinction involve the extinction of *millions* of species at nearly the same geological time. Mass extinctions are not just "more of the same", they are cataclysmic changes in the distribution and numbers of species on Earth.

Within just the past few years, prompted in part by the hypothesis that extinctions may result from large impacts, paleontologists have probed more deeply into the fossil record and have discovered new evidence concerning those times when species disappeared in vast numbers. Their most significant discovery consists of the realization that mass extinctions represent not freakish events but a continuing pattern in the history of life on this planet. Most startling, though not yet confirmed, is the idea that these mass extinctions appear to recur in a periodic cycle.

Fossil Evidence for Mass Extinctions

In thinking about the extinction of species of life on Earth, it is important to bear in mind one key fact: By far the majority of all species that have ever existed on this planet are now extinct. The usual fate for species is total disappearance.

This conclusion rests on an extrapolation from the fossil record. Because we can find living species far more easily than we can discover the fossil traces of vanished ones, we can point today to several tens of millions of species in existence, but only to several hundred thousand extinct species known from the fossil record. However, because paleontologists have a rough idea of how incomplete that record is (from careful studies that focus on one particular time and one particular place), they can estimate the number of species that have existed without their fossils having been catalogued in our museums. This deduction holds that ten to twenty *billion* different species have existed on Earth during the nearly four billion years since life first arose.

Although "only" a few hundred thousand species have been found in the fossil record, even this number proves unwieldy in attempting to chart the record of extinction. For greater simplicity and a more consistent record, paleontologists exploring the history of extinctions have focused on the family level in the classification of living organisms. The family level ranks third from the bottom level (species) of taxonomic classification. Each family includes a number of genera (plural of genus), and each genus contains a number of individual species, and in some cases, subspecies. Families are members, in turn, of orders, which themselves belong to classes, each one part of a phylum. The phyla, finally, belong to one of the broad kingdoms into which biologists classify different types of life. (A useful mnemonic device, known to generations of biology students, gives the initial letters of classification scheme in order from kingdom to species: "King Philip Came Over From Germany Singing.") Human beings, for example, belong to the animal kingdom, the phylum of chordata (and the subphylum of vertebrates), the class of mammals, the order of primates, the family of hominids (of which humans are currently the only representatives), the genus Homo (large-brained hominids capable of speech), and the species Homo sapiens.

Examination of the fossil record at family level, in order to map the extinction of families of marine animals—vertebrate, invertebrate, and protozoan—led the paleobiologists David Raup and John Sepkoski, of the University of Chicago, to a startling discovery during the early 1980s. The fossil record of the past 250 million years shows well over a dozen different peaks in the rate of extinction, four of them major. Impressive though this phenomenon is, the most important implication of the extinction peaks concerns their possible *cyclical* recurrence, for the interval between peaks is nearly constant—as if some "force" returned to the scene of life at regular intervals to produce mass extinctions.

The Thirty-nine Stages

The fossil record for the last 250 million years, since the Permian period that ended the Paleozoic era, is significantly more complete than the fossil record for earlier times. Therefore, in order to make an accurate analysis of changes in the extinction rate, Raup and Sepkoski decided to concentrate on this most recent quarter-billion-year interval. As a result, one of the greatest mass extinctions, at the end of the Devonian period, more than 350 million years ago, does not appear in the fossils that they analyzed—the price to be paid for attempting to restrict the analysis to a time period for which we have an almost complete set of data.

Just as biologists have a taxonomic classification system that relies on divisions and subdivisions, geologists divide time into eras (see page 14), which are further divided into periods. The periods consist of stages, which are the basic units of geological time. Geologists divide the interval between the late Permian period and the Holocene (Recent) period into thirty-nine stages. The oldest of these stages, at the start of the Triassic period, begins 248 million years ago; the youngest, at the end of the Pliocene epoch, comes to a close 2 million years ago. The average duration of each of the thirty-nine stages equals 6.3 million years—more than a moment by human standards, but a finely divided time scale in proportion to the nearly 250 million years covered by all the stages.

Dating the Ages of Rocks

Paleobiologists—those who study the fossil record of life—have made impressive strides in their ability to determine the ages of the rocks that characterize the thiry-nine divisions within the Cenozoic and Mesozoic eras, and of the fossils and rocks laid down during the earlier Paleozoic and Precambrian eras as well. The dating of fossils goes back to the observations of an English surveyor, William Smith, who wanted to predict how easily canals could be cut through layers of rock. Despite his family's disapproval, Smith had spent much of his youth in Oxfordshire collecting common fossils of an extinct sea urchin. (These fossils were used by the local laundrywomen as approximate pound weights and so were called "pound stones.") As his surveying took him through Oxfordshire, Gloucestershire, and Somerset, Smith looked for other fossils, and noticed that different types of fossils appeared in different rock layers. He came to the natural conclusion that the ordering of the layers was the same from place to place. Once Smith had established which fossils were found in which layers, he could predict which layers were beneath the surface by studying the fossils in the surface layer. In 1815, having become known as Strata Smith, he published his *Geological Map of England and Wales, With part of Scotland*; a few years later, Smith sold his collection of fossils to the British Museum, which published a catalog of the fossils under the title *Stratigraphical System of Organized Fossils*. This work laid the foundation for assigning the ordering of ages to different types of fossils, on the basis of the layers, higher or lower, in which they were found.

Similar work throughout Europe (later extended to all the continents and beneath the oceans) allowed paleontologists to discover the worldwide ordering of fossils, layer by layer, with each group of sedimentary rocks (rocks formed underwater) called a system. These systems are the physical basis for the periods of modern geology, each one now subdivided into geological stages. The periods and stages still carry the names of the original localities in which they were studied, though they refer to types of rocks found throughout the world.

Radioactive Dating Techniques

Although geologists knew the *ordering* of the rock systems by the mid-nineteenth century, they could not determine the actual ages of any of the layers. This knowledge had to await the discovery and understanding of radioactivity, the slow transformation of certain types of atomic nuclei into other types. Such transformations, now called *radioactive decay*, affect only some isotopes of certain elements. An element is defined by the number of protons in each atomic nucleus: one proton for hydrogen, two for helium, six for carbon, seven for nitrogen, eight for oxygen, and so on. Isotopes of an element all have the same number of protons, but differing number of neutrons, in each atomic nucleus. Carbon-12, for example, has six neutrons and six protons; carbon-13 has six protons and seven neutrons; carbon-14 has eight neutrons in addition to the six protons. These three isotopes have different abilities to endure indefinitely. Carbon-12 and carbon-13 are not radioactive, but carbon-14 nuclei are: They spontaneously turn into nitrogen-14 isotopes, each with seven protons and seven neutrons, without regard to any outside force that acts upon them.

Although we cannot predict when any one particular carbon-14 nucleus will decay into a nitrogen-14 nucleus, we do know that after 5,560 years one-half of any group of carbon-14 nuclei will have become nitrogen-14 nuclei; after another 5,560 years, one-half of the remaining carbon-14 nuclei will have become nitrogen-14 nuclei; and so on, indefinitely. This means that carbon-14 has a half-life of 5,560 years. At any moment, some of the carbon-14 nuclei are decaying into nitrogen-14 nuclei, so the fraction of carbon-14 nuclei in the sample of material steadily decreases and the nitrogen-14 fraction steadily increases. Since we know the rate of this decay, we can determine the age of a sample of material simply by measuring the ratio of carbon-14 nuclei to the total of all carbon isotopes.

Because bombardment by cosmic rays (fast-moving nuclei and electrons) constantly creates new carbon-14 nuclei on the earth's surface, and because some of these nuclei are assimilated by any organism as it takes in new amounts of carbon, every living creature, ourselves included, has a small fraction of all its carbon nuclei in the form of carbon-14 isotopes. This fraction is the *same* for all living creatures, because the formation of new carbon-14 nuclei by cosmic rays balances the decay of existing carbon-14 nuclei. Death halts the intake of new matter, so from that point the number of carbon-14 nuclei steadily declines. By measuring the amount of carbon-14 that remains in dead organic matter, scientists can determine how many carbon-14 half-lives have elapsed since the organism's demise. For example, two half-lives (11,128 years) leave only one quarter as much carbon-14 as we find in living organisms.

This method works well if only a few, or a few dozen, half-lives have passed, so it has been of tremendous service in reconstructing the archaeological record of the past few tens of thousands of years. Unfortunately, after a hundred thousand years or so, too little carbon-14 remains to be measurable, so the method loses its usefulness.

Luckily, though, other radioactive nuclei decay more slowly. These nuclei can be used to date fossils whose ages range in the millions, rather than in the thousands, of years. The most important of these nuclei are those of potassium-40, which has a half-life of 1.3 billion years, and those of rubidium-87, which has a half-life of 47 billion years. In an approach analogous to carbon-14 dating, geologists measure the amount of the "daughter" nuclei produced by the radioactive decay— argon-40 and strontium-87, respectively. They also estimate how much of these two types of nuclei existed in the rocks when the rocks formed, typically estimating zero for the amount of the daughter isotope. Then, just as in the case of carbon-14 dating, our knowledge of the half-life of the radioactive (naturally decaying) isotopes tells us how much time must have elapsed to produce the parent-daughter ratio measured today.

Dating rocks by radioactive decay works well for those rocks that formed from a molten state—igneous and, in some cases, metamorphic rocks. The heating of such rocks released as gases any argon-40 or strontium-87 within them, so the rocks "began life" with a zero abundance of these "daughter" isotopes. Unfortunately, the method does not work for sedimentary rocks—those that contain the best fossil beds.

To date sedimentary rocks by radioactive decay, a geologist must find the youngest igneous rock below the sedimentary layers, determine the age of that rock, and then assign a slightly lesser age to the sedimentary layers formed above it. Such juxtapositions of sedimentary to igneous rock are not common, but geologists have found enough places where they do occur to establish ages for the thirty-nine systems (and thus for the corresponding stages) that have been laid down over the past 248 million years—ages that are accurate give or take a million years. The *ordering* of the stages, however, remains rock-solid: With the rarest of exceptions, caused by strange geological occurrences, the upper layers are younger than the lower ones, even if the sedimentary layers have been somewhat disturbed since they were first laid down. For this reason geologists will refer to the "upper [younger] Permian" or the "lower [older] Triassic," a reminder that their knowledge comes from actual examination of rock layers piled one on top of another.

Magnetic Reversals and the Ages of Rocks

In the debate over whether mass extinctions recur periodically, and over the possible causes of their recurrence, the basic data to examine consists of the ages of rocks and the fossils that they contain. Without such ages, we cannot even begin to determine the rhythm of mass extinctions. One great aid to geologists in their efforts to find the ages of rocks is their magnetization by the earth's magnetic field. For reasons still unclear, the earth periodically "reverses its polarity," changing the direction of its magnetization by 180 degrees. The record of these magnetic-field reversals, which appear to occur at intervals anywhere from a few thousand to several hundred thousand years, lies preserved in the earth's rocks. Although the earth has a relatively weak magnetic field (Jupiter's, for example, is a hundred times stronger), it can magnetize the minerals in rocks. When an igneous rock cools from a high temperature, it passes through what is known as the Curie-point temperature (different for each variety of rock), at which the earth's magnetic field is "frozen" into the rock. At lower temperatures, the rock resists changes in magnetization much more strongly than it does above the Curie-point temperature. If the rock is left relatively undisturbed, it will preserve as residual magnetism the record of the earth's magnetic field at the time the rock cooled.

This "fossil magnetism" has proved to be a great boon to rock daters: By studying rocks from all corners of the world, they can match the patterns of magnetic-field reversals from one location to another by examining rocks of the same age. In a similar way, radioactive decay allows geologists to date many rocks. Matching these rocks to other rocks through the record of residual magnetism then allows them to date *any* rock and the fossils that it contains.

Because only igneous rocks have residual magnetism, the exact dates of all rock layers throughout the world, and of the geological stages that they represent, remain more uncertain than we would like, though far more accurately established than a few dozen years ago. For time intervals in which no rocks can be accurately dated, geologists rely on intelligent interpolations between the ages of rocks they have dated with some confidence. In this way, Brian Harland, of Cambridge University, has established the most widely acepted and (it is to be hoped) the most accurate chronology for the rocks of the past 600 million years. The Harland chronology, which we shall use to examine the question of whether extinctions recur periodically, differs slightly from the previous chronology. This helps to remind us that the ages assigned to the fossil record contain errors of a million years or so when we look backward to the Cretaceous-Tertiary boundary. The ages assigned to older rocks are subject to significantly greater errors, easily reaching 10 million years—and possibly 20 million—by the time that we reach the Permian-Triassic boundary, generally established at 248 million years ago. These larger errors increase in a cumulative manner, since we work backward in time—downward through the rock strata—dating each layer by the previous ones. We must bear these errors in mind when we direct our attention to the heart of the Shiva theory: the claim that mass extinctions are periodic.

Do Extinctions Recur in a Periodic Cycle?

Raup and Sepkoski were not the first to look for a periodic cycle in the extinction data, but they had a more accurate set of data to examine than their predecessors. Suggestions that extinctions have occurred in cycles originally were made in 1970 by the geologists Craig Hatfield and Mark Camp of the University of Toledo, and in 1977 by Alfred Fischer and Michael Arthur of Princeton University. But only when Raup and Sepkoski presented their detailed analysis of the fossil family record did the world of science sit up and take notice.

When Raup and Sepkoski drew their graph showing the percentages of families going extinct during each geological stage since the Late Permian period, they found an intriguing combination of peaks and valleys. Raup and Sepkoski defined a peak in the graph as any stage in which a larger percentage of families become extinct than in the two surrounding stages. Using this definition, they found twelve peaks in the data for the thirty-nine geological stages. Since that time, they have discarded four of the peaks as too small to be significant, so their results now show eight identifiable peaks. Raup and Sepkoski engaged in mathematical analysis to see whether the times at which these peaks have occurred show any periodic—regularly recurring—behavior.

Statistical analysis rarely convinces everyone, but the standard test performed on the data assembled by Raup and Sepkoski reveal periodic behavior at a high level of confidence. A common statistical test, the "Monte Carlo calculation," examines a *random* set of numbers to see how often such a set exhibits behavior like that of the actual data—in this case, periodicity in the extinction peaks—to the same degree. By

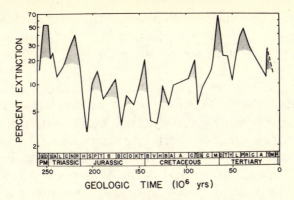

Figure 4—The graph made by David Raup and J. John Sepkoski, Jr., for their paper in *Proceedings of the National Academy of Sciences* shows the extinction of families, in percent, from one geological stage to the next. The stages are represented by single-letter abbreviations above the names of the periods that they subdivide (e.g., E stands for Eocene). Four of the peaks stand out dramatically, and three or four others seem significant. The time scale has an accuracy that varies from plus or minus one million to two million years in the Tertiary period to plus or minus ten million to twenty million years in the Triassic period. (Courtesy of Prof. David Raup.)

performing this test with hundreds or thousands of random data sets, one can see how many times in a thousand the random data show as strong an example of the claimed behavior as the actual data do. If, for example, the random data agree with the actual data four hundred times in a thousand, then the actual pattern almost certainly represents nothing more than the randomness of the world in one of its many guises.

According to the tests performed by Raup and Sepkoski, we can have confidence at a level greater than 90 percent that the extinction rate shows a cyclical pattern, with a period of 26 million years between successive extinction peaks. A 90-percent level of confidence means that only once in every ten tries does a series of random numbers match the periodic cycle equally well. Raup and Sepkoski found weaker evidence for a

30-million-year cycle in their data; they concluded that the extinction data *do* exhibit periodic behavior, with an interval between peaks somewhere from 25 to 27 million years in length.

When you look at the graph made by Raup and Sepkoski, it is clear that the fit between peaks in the actual data and a hypothetical recurring cycle of extinction with a period of 26 million years is good but certainly not perfect. The four largest of these peaks occur at the Permian-Triassic boundary, 248 million years ago; at the end of the Norian stage in the Triassic period, 219 million years ago; at the Cretaceous-Tertiary boundary, 65 million years ago; and at the Eocene-Oligocene boundary at the end of the Tertiary period, 38 million years ago.

A perfect 26-million-year cycle postulates ten extinction peaks within the last 253 million years. Of these ten proposed peaks, seven match quite well with actual extinction peaks found by Raup and Sepkoski. Fixing the most recent such predicted peak at 13 million years ago, these seven good fits between the actual and theoretical cycles occur at the Eocene-Oligocene boundary, 38 million years ago; at the Cretaceous-Tertiary boundary, 65 million years ago; at the boundary of the Cenomanian and Turonian stages, 91 million years ago; at the boundary of the Jurassic and Cretaceous periods, 144 million years ago; at the end of the Pliensbachian stage of the Jurassic period, 194 million years ago; at the end of the Norian period, 219 million years ago; and at the end of the Permian period, 248 million years ago. The perfect cycle requires that these peaks should have occurred 39, 65, 91, 143, 195, 221, and 247 million years ago. These estimates do, in fact, fall within one or two million years of the actual peaks recorded, and thus within the margin of error in dating the rocks that give us these times. These seven good fits include all four major peaks in the extinction data. The lesser peaks do not fit the cycle so well, but then again, they may not be statistically significant. As we shall see, the 26-million-year cycle provides a marginally better fit to the extinction data than the competing 30-million-year cycle of the galactic oscillation theory.

Statistical analysis confirms that the eight major peaks in the extinction data fit quite well to the idealized cycle of 26 million years. Raup and Sepkoski's paper announcing these results, published in *Proceedings of the National Academy of Sciences* in February 1984, may have ushered in a new era in thinking about extinctions on Earth. *Some* extinctions occur during every geological stage; the extinction rate rarely falls below 5 percent, and never below 2.5 percent, for all of the thirty-nine geological stages that span the past 250 million years. But the peak extinctions found by Raup and Sepkoski average close to 20 percent. The four largest of their peaks rise to 40–50 percent, and even to nearly 70 percent of all *families*. The latter figure applies to the Permian-Triassic extinction and to the Cretaceous-Tertiary extinction that started things off for Asaro, Michel, and the Alvarezes. Think of an event that removed not simply 70 percent of the *species* then present but 70 percent of the *families* of living creatures! Then reflect that this appears to be merely the most prominent of cyclical mass extinctions, recurring every 26 million years, that characterize at least the past 250 million years of life on Earth. You may then be ready to ask what could possibly cause such recurrences—provided, of course, that you believe the data and analysis of Raup and Sepkoski.

The fact is, however, that not everyone agrees that the data so carefully assembled by Raup and Sepkoski exhibit a 26-million-year cycle. The competing period to fit the data lies close to 30 million years, and is dear to those who support the galactic-oscillation hypothesis of mass extinctions, which we shall examine in chapter 4. In order to understand the arguments and counterarguments a bit better, it is useful to examine some more familiar numbers in order to get a better feeling for what arguments for periodicity in a series of numbers can

Observed Extinction Peaks and Predictions of Cyclical Theories

Period and Stage	Time before present (in million years) of measured peak, based on Harland chronology	Time before present of closest peak predicted by 26-million-year cycle (Nemesis theory)	Time before present of closest peak predicted by 30-million-year cycle (galactic oscillation theory)
TERTIARY			
Middle			
Miocene	11.3	13	5
Late Eocene	38	39	35
CRETACEOUS			
Maestrichtian	65	65	65
Cenomanian	91	91	95
Hauterivian	125	117	125
JURASSIC			
Tithonian	144	143	155
Callovian	163	169	155
Bajocian	175	169	185
Pliensbachian	194	195	185
TRIASSIC			
Norian	219	221	215
Olenekian	243	247	245
PERMIAN			
Dzhulfian	248	247	245

From "Periodicity of Extinctions in the Geologic Past," by David M. Raup and J. John Sepkoski, Jr., published in *Proceedings of the National Academy of Sciences*, February 1984, vol. 81, pp. 801–805.

be like. It is a sad fact that most citizens have no love for numbers, and far from loving them for themselves, or for what they might do for us, these people regard numbers as demons they confronted during childhood with only modest success, and can thankfully have only minimal contact with during their adult life. Mathematicians are thought of as odd individuals who love numbers. This is a slur on numbers as well as on mathematicians, most of whom deal with numbers no more than anyone else, though a few of them are indeed involved in number theory—the search for the magic of pure numbers. Other scientists do use numbers every day and "argue the numbers" routinely, quoting how many of this or that do or do not support their favorite hypotheses. The argument over whether or not we can demonstrate periodic behavior, and of what kind, in the timing of mass extinctions furnishes a fine example of this sort of debate. To get a better handle on it, we may consider, by way of example, a famous series of numbers.

Is This Series Periodic?

Think of the following series of eleven numbers, imagining them, if you like, as the dates of mass extinctions in millions of years before the present:

4   14   23   34   42   50   59   72   81   86   96

Can you say with confidence whether this series exhibits periodic behavior—that is, whether it shows an overall trend to increase by a fixed amount, perhaps with modest variations in this amount?

A scientist would insist first of all on being told how accurate these numbers are. If, for instance, each of these numbers represented a date of mass extinction, given in millions of years, and each date were known only to within 8 or 9 million years, there would be not much point in trying to see whether the numbers given above show periodic behavior, since many of the successive dates might represent the same time. If, on the other hand, the dates were known to be accurate within 2 million years, then the data would be precise enough to make looking for periodicity a natural and sensible undertaking.

Let us imagine that the latter case holds true for the numbers above. Now, various types of statistical manipulation can be performed on the number series, of which one of the simplest is to note the differences between its successive members:

$$10 \quad 9 \quad 11 \quad 8 \quad 8 \quad 9 \quad 13 \quad 9 \quad 5 \quad 10$$

These certainly differ, but not tremendously. They suggest, among other possibilities, periodic behavior with an average increase of 9.2 between successive terms. If we round off the average increase to 9, and allow ourselves to begin the idealized series with the number 5 (instead of 4, as in the actual series), then we obtain the idealized series:

$$5 \quad 14 \quad 23 \quad 32 \quad 41 \quad 50 \quad 59 \quad 68 \quad 77 \quad 86 \quad 95$$

This would fit rather well to the actual series. Except for 68 and 77 in the idealized series (as compared to 72 and 81 in the actual series), the misfits are each no greater than 2. Hence, in this case, we can make a plausible argument for periodic behavior with a cycle of 9. Statistical analysis would show that these data exhibit periodicity about as strongly as the Raup-Sepkoski data on mass extinctions.

A common way to measure the deviation of an actual series of data from an idealized, perfectly periodic series is to take the differences between each member of the actual series and the value predicted by the idealized series. These differences are squared (thus positive and negative differences contribute equally, rather than canceling one another out) and added together. Their sum gives the measure of deviation for a series of a given length. This kind of calculation shows that about the same deviation exists between the actual and idealized series created in our example as does between the actual mass-extinction data and the idealized, 26-million-year periodic cycle claimed to fit that data.

Our original series is in fact familiar to almost every New Yorker over the age of fifteen, since it represents the subway stops (street numbers) on the Eighth Avenue line. A little reflection on how subways are designed leaves us unsurprised to find that the spacing between stops is roughly constant throughout the most densely packed parts of Manhattan, and that this spacing averages close to 10 blocks.

Notice, however, that it would not take a great change in the data to find cyclical behavior with a significantly different period. Consider this series of eleven numbers:

14   18   23   28   34   42   50   59   66   72   79

Seven of the eleven match the numbers in the original series, but now the intervals between successive members of the series are:

4   5   5   6   8   8   9   7   6   7

These intervals average 6.5, so if we start our idealized series with 14, we find the series:

14   20.5   27   33.5   40   46.5   53   59.5   66   72.5   79

The fit is not quite as good as it was for the cycle-of-9 series, but it is nevertheless statistically significant. Here we have the stops along the Broadway subway line; many of them match the numbers of the Eight Avenue subway, but they occur with a period of about 6½ blocks rather than 9 blocks—a tribute to the age of the Broadway line, designed for a slower-moving era.

Suppose that the stops at 28th Street and 66th Street had been left out of the Broadway line, or that we failed to notice them as we sped past. We would then try to analyze the series:

14 18 23 34 42 50 59 72 79

This series has intervals of:

4 5 11 8 8 9 13 7

The differences here average to 8.1, so if we use an interval of 8, our idealized series, adjusted to start with 16 instead of 14, becomes:

16 24 32 40 48 56 64 72 80

This idealized series does not fit the data as well as the $6\frac{1}{2}$-block interval fits the actual Broadway line, which in turn provides a poorer fit than that between the 9-block interval and the Eighth Avenue subway line. The joys of statistics lie in refining these statements, reducing them to numbers, and allowing scientists to argue about the significance of the numbers.

Which Period Appears in the Mass Extinction Data?

If Raup and Sepkoski's data failed to reveal one or more actual peaks in the extinction data, then they would have derived too long a period between mass extinctions; if they had included one or more spurious peaks, they would have underestimated this period. Furthermore, as is intuitively obvious, if they failed to include some real peaks, or did include some spurious ones, they would arrive at a period in which they ought to have less confidence than if they knew they were analyzing all the real peaks and none of the spurious ones. Critics of the derived periodicity point out that the Raup-Sepkoski data are so incomplete that attempts to determine whether or not they show periodic behavior resemble attempts to uncover the periodicity of subway stations without being sure that one can tell the stations from the tunnel in between.

As the table of extinction dates shows, we can also try to fit the mass-extinction data with a cycle of 30 million years rather than 26 million. This seemingly small difference in proposed periods has tremendous repercussions in the debate over the *cause* of mass extinctions. It turns out that a 30-million-year cycle could (almost) be explained as the result of the sun's motion in the Milky Way (the galactic-oscillation theory). According to this theory, as the sun passes through the Milky Way's median plane (at intervals of a bit more than 30 million years), comet showers arise from close encounters with interstellar clouds of gas and dust. The opposing Nemesis theory proposes a solar companion star, moving around the sun with a period of 26 million years, to explain comet showers and mass extinctions. The 26-million-year cycle is too short to be explained by the galactic-oscillation theory. *Therefore if the 26-million-year and not the 30-million-year period proves valid, the galactic-oscillation theory seems doomed, and the Nemesis hypothesis becomes more likely.* On the sidelines of this dispute, occasionally uttering caustic remarks, we find those who doubt that the data prove any periodicity at a level of confidence worthy of attention.

A Closer Look at the Extinction Peaks

Suppose we look at just the four largest extinction peaks that appear in Raup and Sepkoski's data. These peaks occur at times (in millions of years before the present) of:

$$38 \quad 65 \quad 219 \quad 248$$

and the intervals between them (in millions of years) are:

$$27 \quad 154 \quad 29$$

Raup and Sepkoski's fitting of a 26-million-year period nearly matches the first gap, between 38 and 65 million years, but does more poorly with the last, between 219 and 248 million years, and it puts six cycles (156 million years) between the second and third peaks. The alternative possibility, the 30-million-year cycle, fits the four large peaks just about as well, but with only five cycles (150 million years) between the second and third peaks, rather than the six predicted by the 26-million-year period. Now suppose that we also include the two next highest peaks in the extinction data. This gives us a series of times before the present equal to:

$$38 \quad 65 \quad 91 \quad 144 \quad 219 \quad 248$$

The differences now become:

$$27 \quad 26 \quad 53 \quad 75 \quad 29$$

This pattern seems to favor the 26-million-year cycle, since the first two intervals are close to 26, the next interval is just about double, and the third just about triple. The full idealized cycle would show the times:

$$13 \quad 39 \quad 65 \quad 91 \quad 117 \quad 143 \quad 169 \quad 195 \quad 221 \quad 247$$

These times may then be compared with the times of the actual peaks in the Raup-Sepkoski data, which are:

(11) 38 65 91 (125) 144 (163) (175) 194 219 (243) 248

The dates in parentheses represent the less prominent extinction peaks, which may not be actual peaks, and which we might do best to ignore when we imagine what really happened on Earth.

However, more sophisticated analysis, carried out by Richard Stothers and Michael Rampino, of NASA's Goddard Space Institute in New York City, shows that we cannot entirely rule out a 30- or 31-million-year cycle in the data assembled by Raup and Sepkoski. To make such a cycle fit, though, we must place the most recent peak at only a few million years ago, rather than at 13 million years ago (as we must do in order to fit the data with a cycle of 26 million years). Then the dates of mass extinctions predicted by a 30-million-year cycle become:

5 35 65 95 125 155 185 215 245

Since the four most significant extinctions occurred about 38, 65, 219, and 248 million years ago, we can see that the fit is not all that bad, even though the most recent peak indicated by the cycle does not appear in the extinction data. This absence of the most recent peak also holds true for the 26-million-year cycle, although there is some evidence for a small extinction peak 11.3 million years ago. The 30-million-year cycle also misses the major peaks at 144 and 194 million years ago, and nearly misses the Cenomanian-Turonian extinction 91 million years ago.

Rampino and Stothers also claim statistical significance for a 31-million-year period. We can attempt to fit the data with such a cycle, placing the first peak 3 million years ago, as follows:

3 34 65 96 127 158 189 220 251

This series "hits" the actual peaks at 65, 219, and 248 million years ago (with some allowance for error in dating the peaks) but misses the important peaks at 38, 91, 144, and 193 million years ago by slightly larger margins than can be said to lie within the limits of error for the first two of these four extinction peaks. The 30-million-year cycle does somewhat better than the 31-million-year cycle in matching the most important peaks, but as Rampino and Stothers point out, both must be considered acceptable until disproved in scientific debate.

The bland pronouncement that a debate exists between advocates of different periodicities in the allegedly cyclical extinctions hardly describes the state of tension that exists between those who claim to have demonstrated the existence of a 26-million-year cycle, with its most recent peak about 13 million years ago, and those who find a 30- or 31-million-year cycle, with its most recent peak 2 to 6 million years ago. One might think that this argument would be considered minor, and that the important matter would be that convincing evidence of a recurrent cycle—with a period of either about 26 million or about 31 million years—emerges from the Raup and Sepkoski data. But as we shall see, such an easygoing viewpoint overlooks the importance of the two different estimates to theories that argue *two different causes* for the cycle of mass extinctions.

Furthermore, the two proposed cycles differ markedly in their prediction of the *next* peak in extinction, though both imply that we need not hold our breath. According to supporters of the 26-million-year cycle, we are now just about halfway between extinction peaks, and so should expect the next peak in 13 million years. The 30-million-year cycle, in contrast, has us only "recently" emerged from a peak and due for the next one only after another 25 million years have passed, almost twice as long a breathing space as the 26-million-year cycle would provide.

The Record of Impact Craters

The analysis of the fossil record made by Raup and Sepkoski during the early 1980s represents an impressive piece of research, but by itself this research says nothing about the possible cause of cyclical mass extinctions. Considering that in 1980 the Alvarezes, Asaro, and Michel had already proposed the asteroid-impact hypothesis to explain the extinctions at the Cretaceous-Tertiary boundary, it is not surprising that scientists stood ready to examine whether any further evidence for impacts on Earth could be tied to the new evidence from the fossil record. Geologists and physicists looked, and found, not perfect evidence—not even convincing evidence to many scientists—but evidence that at least suggests a cyclical pattern in the record of the earth's *impact craters* similar to that of mass extinctions.

Geologists can assign fairly accurate dates to most of these craters by examining the rocks at the site that were melted by the impact, using the radioactive decay of potassium isotopes into argon isotopes as a "clock" for dating once-molten rocks. The melting through impact in effect restarted the clock, because it allowed the gaseous argon previously formed by decay and trapped within the rocks to escape. Measurement of the amount of "daughter" argon formed through radioactive decay then dates the rock to the time that it was last in a molten state.

In 1982, the geologist Richard Grieve, of the Canadian Department of Energy, Mines, and Resources, searched the published literature on crater ages and compiled a list of all known impact craters younger than 600 million years. Although only upper limits could be assigned to the ages of some craters, most have ages known to within plus or minus two to five million years. This imprecise dating of impact craters—which allows for a 10- or even 20-million-year error for some of the older craters—means that more than one claim can seem

Impact Craters With Ages From 5 to 250 Million Years and With Diameters of at Least 5 Kilometers

Crater Diameter (Kilometers) and Location	Age (Millions of Years)	Age as Revised by Alvarez and Muller	Closest Peak in Time of Proposed 28.4-Million-Year Cycle of Impacts	Closest Peak in Time of Proposed 32-Million-Year Cycle of Impacts
10 Karla, U.S.S.R.		7 ± 4	13	3
20 Canada		13 ± 11	13	3
24 Ries, Germany	14.8 ± 0.7		13	3
28 Labrador	38 ± 4		41.4	35
8.5 Ontario	37 ± 2		41.4	35
100 Siberia	39 ± 9		41.4	35
14 Finland	77 ± 4	78 ± 2	69.8	67
25 Alberta	95 ± 7		98.2	99
25 Ukraine	100 ± 5		98.2	99
17 Logoisk, U.S.S.R.	100 ± 20		98.2	99
5 Sweden	118 ± 2	119 ± 2	126.6	131
22 Australia	130 ± 6	133 ± 6	126.6	131
23 France	160 ± 5		155	163
80	183 ± 3		183.4	195

Crater Diameter (Kilometers) and Location	Age (Millions of Years)	Age as Revised by Alvarez and Muller	Closest Peak in Time of Proposed 28.4-Million-Year Cycle of Impacts	Closest Peak in Time of Proposed 32-Million-Year Cycle of Impacts
U.S.S.R.				
15		185 ± 10	183.4	195
Ukraine				
70	210 ± 4	214 ± 3	211.8	227
Quebec				

Note: The first column of ages is from the compilation by Grieve, and was used by Alvarez and Muller in their statistical analysis. The revised age estimates were made by Alvarez and Muller, using new data on the rates of radioactive decay of certain isotopes.

valid when it comes to estimating the period of the cyclical recurrence of impact craters on Earth.

Like the data on mass extinctions, Grieve's list has been the focus of intense debate among scientists who have searched it to see whether the crater ages show periodicity. Everyone agrees that not all the craters on Grieve's list should be included in any statistical analysis, but there is much disagreement as to *which* craters are to be excluded.

First, for twenty-two of the craters on Grieve's list we have only upper limits for their ages, rather than estimates of the ages themselves. All agree that such craters should not be subjected to statistical analysis. Second, the remaining craters on the list may not be a representative sampling, because erosion gradually removes the traces of impact craters, so that many older, small craters have disappeared completely. The craters of today therefore do not provide an accurate record of the past; older craters, especially small ones, tend to be missing from the list despite geologists' efforts to find them, simply because they are too severely eroded to be discernible. Thus it is difficult to compile what we would like to have—a list of all the impact craters that have ever been made, rather than those whose remnants we can discern.

Berkeley's Walter Alvarez and Richard Muller, and Michael Rampino and Richard Stothers of NASA's Goddard Institute of Space Studies separately imposed criteria designed to reduce the "bias" inevitably contained in Grieve's list of craters, and then analyzed their restricted lists to see whether the crater ages show periodicity. And indeed the Berkeley and New York scientists found a cyclical pattern, but once again the problem is that their findings differ. Alvarez and Muller found a period of 28.4 million years, but Rampino and Stothers derived a period of 32 million years. In other words, both the Berkeley and the NASA scientists found the period that nearly corresponds to each group's choice of the period between mass extinctions.

Science differs, however, from most other human endeavors in that the data remain the data, and each of us can examine them to reach our own conclusions. Neither analysis of the data on crater ages is fundamentally incorrect; all of these scientists know how to make mathematical calculations. The results that the two groups reached were different because they did not analyze quite the same set of data. A quick look at the "facts" will show how the two groups of scientists derived two different periods from the record of crater ages.

The 28.4-Million Year Period

Alvarez and Muller analyzed only the largest impact craters—those with diameters of at least 10 kilometers—eliminating all craters with assigned ages of less than 5 million years. In making these choices, they sought to compensate for any bias introduced by the missing older, smaller craters. Rampino and

Stothers could claim that Alvarez and Muller's attempt to compensate for this effect threw out the periodic baby with the bias-laden bath, for Grieve's list of eighty-eight craters with ages from less than 1 million to 600 million years contains twelve craters younger than 5 million years—13.6 percent of the total number in less than 1 percent of the total time interval.

These youngest craters form the largest peak in the crater-age data, and (if included) provide support for the finding of periodic behavior with its most recent peak only a few million years ago. In their analysis, Alvarez and Muller also discarded those craters with ages greater than 250 million years, because they wanted to check any periodicity they might find against the periodicity in the extinction rate found by Raup and Sepkoski, whose analysis covered the past 250 million years. Finally, they included only those craters whose uncertainty in age is less than 20 million years. This represents an entirely reasonable refusal to look for possible periodic behavior among crater ages whose range of uncertainty almost equals the possible period.

The criteria set by Alvarez and Muller left just thirteen craters (from Grieve's compiled list of eighty eight), with ages that range from 7 million to 214 million years. The oldest, with an age close to 214 million years, is the well-known Manicouagan Crater in northern Quebec, 70 kilometers in diameter, highly eroded but still easily visible in a satellite photograph. Three of the thirteen craters have ages close to 38 million years: one in Labrador, one in Ontario, and one in Siberia. Two others have ages close to 100 million years, one in the Ukraine and one in Asiatic Russia. Another crater, in West Germany, has a well-defined age of 14.8 ± 0.7 million years, where the plus or minus indicates the range within which the true age is likely to fall at a confidence level of 95 percent.

When Alvarez and Muller subjected these ages to their statistical tests, attempting to see what period, if any, for a cyclical recurrence in the crater ages would make the cyclical

Figure 5—The Manicouagan Lakes Crater in northern Quebec, 70 kilometers in diameter, has been dated to the late Triassic period, about 214 million years ago. In another 200 million years, erosion will make it difficult to find the outlines of the crater, which is now believed to have been caused by an impact, as shown by the complex, multiringed structure discernible around the main crater. (NASA ERTS photograph.)

pattern best fit the data, they found a good correlation between the data and a cycle whose period equals 28.4 million years. Equally important, the first peak in the derived cycle of crater ages falls 13 million years ago, just in step with the cycle of the extinction rate found by Raup and Sepkoski. (Notice, however, that there is *no* crater with an age of 13 million years; we are fitting a curve to a series of points, and the closest age in this fit is 14.8 million years.) The second peak in the derived cycle, looking backward from the present, occurs 41.4 million years ago, reasonably close to the time of the Eocene-Oligocene boundary (38 million years ago). Representing this peak in reality we find three craters with ages of 37, 38, and 39 million years, and possible errors in age of 2, 4, and 9 million years respectively. The third peak in the idealized cycle, which occurs 69.8 million years ago, has no representatives in the short list of eleven craters; we may also notice that it occurs almost 5 million years before the currently assigned age of the Cretaceous-Tertiary extinction. The fourth peak backward occurs 98.2 million years ago in the derived cycle; we find two craters with ages of 100 million years on the list, one age with an error of ± 5 million years, the other ± 20 million years. The fifth peak is predicted at 126.6 million years; there is one crater with an age of 130 ± 6 million years. The predicted sixth peak backward lies 155 million years ago; the Rochechouart Crater in France has an age of 160 ± 5 million years. The seventh peak occurs 183.4 million years ago; a crater in Russia has an age of 183 ± 3 million years. Finally, the eighth peak backward lies at 211.8 million years ago, and the Manicouagan Crater, as we have seen, has an age of 214 ± 3 million years.

A Far From Perfect Fit

Hence we might conclude that the periodicity found by Alvarez and Muller in the crater ages is far from a perfect match with the periodicity found by Raup and Sepkoski in the rate of extinction. Indeed, Alvarez and Muller are the first to admit that a 28.4-million-year period for the crater ages is not the same as a 26-million-year period for the extinctions. They point out, however, that the dates determined for seven of the eight most significant extinctions found by Raup and Sepkoski—those for which more than 20 percent of the families under study disappeared—fit the 28.4-million-year cycle within the error limits assigned to them. Because the extinction and impact-crater periods differ by only 2.4 million years, the first four peaks in both sets of dates—looking backward in time—can almost coincide, but the 2.4-million-year difference does accumulate between the crater-age and extinction-rate cycles. Then, during the next hundred million years backward in time, a slippage by one complete cycle must occur as we look back to the Norian extinction 219 million years ago, and to the Permian-Triassic extinction 248 million years ago. In other words, the error limits in the data allow for the possibility of a 26-million-year cycle of extinction and a 28.4-million-year cycle of cratering to coexist, simply because we do not know the times of extinction or the ages of craters with complete accuracy.

To test whether the derived period still emerges from the data if smaller craters are included, Alvarez and Muller studied Grieve's list again, this time including in their analysis craters with a minimum diameter anywhere between 0 and 20 kilometers. They found a 28.4-million-year cycle for all minimum sizes, though the cycle was most statistically significant when the minimum crater diameter for inclusion was 5 kilometers. Similarly, Alvarez and Muller found that including craters younger than 5 million years or older than 250 million years did not change the period they found in the data, though it did make that period less statistically significant.

The 32-Million-Year Period

The analysis made by Rampino and Stothers, which finds a period of 32 million years in the impact crater ages, differs most sharply from that of Alvarez and Muller in its inclusion of craters younger than 5 million years (but older than 1 million years).

Both groups of scientists sought to eliminate the bias created by the absence of many older, eroded craters, but each group made a different decision about the cut-off age at which to discard the youngest craters by way of compensation. Rampino and Stothers would claim that Alvarez and Muller discarded too much real data because of excessive concern about the bias created by the youngest craters.

Rampino and Stothers analyzed the ages of sixty-five impact craters on Grieve's list with ages as great as 600 million years. In addition to including those craters with ages greater than 1 million years (not greater than 5 million years, as in Alvarez and Muller's first analysis), they did not restrict themselves to the large impact craters, though they did eliminate those craters whose ages are given only as upper limits. The ages of the sixty-five craters suggest that impacts occur cyclically with a period of 32 million years.

When Rampino and Stothers analyzed the ages of the forty-one craters with ages between 1 and 250 million years, they again found a significant periodicity. This periodicity has its first peak, looking backard from the present, 10 million years ago, and shows a period of 31 million years. Later, these scientists analyzed the data for the craters with diameters greater than 10 kilometers, over a range of ages from a few million to 365 million years. These large craters have long survival times on the earth's surface, so it makes sense to include even the most recent. For these craters' ages, Rampino and Stothers found evidence of a cycle with a period of 32 million years. When they analyzed the ages of the twenty-two craters of all sizes dated to the Paleozoic era (more than 250 million years old), they derived a period of 33 million years.

Recently, Eugene Shoemaker, of the California Institute of Technology and the United States Geological Survey, reanalyzed the data used by Rampino and Stothers. Shoemaker suggests a period of approximately 31 million years, much like the Rampino-Stothers result, but he also finds that the best fit places the most recent peak only a few million years ago, when many of the youngest craters (those between 1 and 5 million years old) were created—craters that were eliminated from Alvarez and Muller's analysis to compensate for vanished older craters. The last column in the table of crater ages presents the adjustment of the Rampino-Stothers analysis, to show that a 32-million-year cycle, with its first peak 3 million years in the past, can also fit the data fairly well, with some noticeable "misses" between the idealized cycle and the ages of large craters.

Shoemaker's analysis confirms that the difference between the results found by Rampino and Stothers and those found by Alvarez and Muller is not really due to the cut-off in crater sizes, or to the restriction to the past 250 million, rather than 365 million, years. What does make a difference is the inclusion of the youngest craters, which comprise more than 10 percent of Grieve's list. These craters strongly influence the statistical analysis by their relatively large numbers, so that the most recent burst of cratering "demands" a peak, and this peak must be about 32 million years prior to the peak at 38 million years ago, which all analyses see clearly in Grieve's list. In fact, the crater statistics put the first peak backward in time at only about 3 million years ago, as we would expect from the abundance of young craters, rather than at 10 million years ago, as derived by Rampino and Stothers.

Each approach to rejecting *some* of the craters on the list has merit—and flaws. Unfortunately, the record of impact craters *on Earth* is just not good enough to enable us to discriminate between a 28.4-million-year period, on the one hand, and a 31- or 32-million year period on the other. Ironically, the answer smiles down on geologists (and the rest of us) on many a night—from the airless, waterless moon, with its well-preserved record of craters, unaffected by any erosion. If we could travel across the lunar surface, collecting rock samples at dozens of craters for radioactive dating, we should soon know whether a periodicity exists in lunar impact-crater ages, and with what period. So far, rock samples returned from the moon have come from only a few locations, almost all of them much older (3 to 4 billion years) than the ages of impact craters on Earth. To resolve the riddle of crater ages would require more energy in lunar rock-collecting than humanity has yet demonstrated, and yet this might occur if we ever return to the moon for a more complete investigation.

Searches for Other Cycles

In addition to their analysis of crater ages, Rampino and Stothers looked for periods in other phenomena preserved in the geological record: times of low sea level, discontinuities in the plate-tectonic spreading of the sea floor, reversals in the geomagnetic field, and other evidence of changes in tectonic activity. According to their research, these phenomena all show evidence of cyclical changes with a period between 33 and 35 million years. However, the *peaks* of the various periods do not coincide in time: The most recent peak occurred 13 million years ago for the discontinuities in sea-floor spreading, but only 2 million years ago for the geomagnetic reversals. If we put this last point aside as not proved, because of the uncertainties in the ages we assign to the various events, we confront the root of the argument: Rampino and Stothers derive a period for the impact-crater ages that simply does not match the 26 million years found by Raup and Sepkoski for the extinction rate.

As we have seen, Rampino and Stothers have an answer to this problem: Their analysis of the extinction data used by Raup and Sepkoski produces a period not of 26 million years but of 30 or 31 million years. Hence both teams—New York and Berkeley—arrive at coherent though mutually contradictory results. Rampino and Stothers find that mass extinctions occur with a period that nearly matches the periodicity of 32 million years that they find in the ages of impact craters. Alvarez and Muller agree with Raup and Sepkoski that mass extinctions show a 26-million-year period, and they find a 28.4-million-year period in the ages of impact craters, which they can fit to the cycles of mass extinction if they slip one cycle in the interval from 120 million to 180 million years ago.

What Can We Conclude About Impact-Crater Ages?

We may be tempted to abandon all hope of knowing whom to believe in the statistics of crater ages and extinction rates, if the best-known scientists in the field cannot agree. But this would be premature; the more accurate conclusion would be that we must deal with a *double* uncertainty in our attempt to look for periodicity in the ages of impact craters. We face one kind of uncertainty in deciding just which data to analyze—all extinction peaks or the major ones, the ages of all craters or of the largest ones. This uncertainty is separate from the uncertainty that arises from the imprecise dating of the times of mass extinctions and of crater ages themselves.

No one claims that either set of data provides a perfect match to a periodic cycle, but both teams of scientists do claim, and with mathematical correctness, that the periods derived from their analyses are highly significant in statistical terms. On one point, Alvarez and Muller firmly agree with Rampino and Stothers: The two teams cannot both be right. During the past 250 million years, many more impact craters than average may have been created every 28 million years, or every 32 million years, but not both. The existence of this controversy serves to alert the novice that neither case has been proved. Only those with some mathematical sophistication, plus an understanding of which impact craters are to be included for analysis, are likely to reach a conclusion with the confidence that is shared by Alvarez and Muller, and Rampino and Stothers. The rest of us can hope that more accurate determination of crater ages may help establish whether one of the two proposed cycles has been demonstrated with sufficient certainty to convince most scientists working in this field, or whether the evidence from crater ages, even though suggestive, cannot be firmly tied to any periodic cycle, and in particular to a 26-million-year or 33-million-year cycle.

REBUTTAL: Who Believes Statistical Arguments?

Scientists who have looked briefly at the arguments made by Alvarez and Muller, by Rampino and Stothers, and by Shoemaker have been known to snort in disapproval of trying to build elaborate theories on the statistical analysis of relatively uncertain data on extinctions and crater ages. Wait and see, these scientists say: New data will disprove both periodic hypotheses. After all, the fewer data points one has, the more easily one can fit a periodic cycle to them. Perhaps all those who find periodic behavior in the data resemble those who would conclude from a ride on the Eighth Avenue line that all of New York's subway stops are spaced nine blocks apart, or would derive a universal spacing of six blocks based on a subway trip up Broadway. Such conclusions would be highly unjustified, though not so completely wrong as to be laughable; after all, the Broadway and Eighth Avenue lines were built on much the same principles as the rest of the subway system.

But, the skeptics would say, the actual data on mass extinctions and crater ages are so uncertain that our model here is like a subway ride on which we cannot determine just where the subway stops (mass extinctions or significant impact craters) are. Hence any period that we derive from the data remains highly suspect, because we can't be sure that our data present a complete record of extinctions on Earth. In fact, a skeptic would say, only four of the mass extinctions in the data are truly impressive: those that occurred 38, 65, 219, and 248 million years ago. To conclude from these that mass extinctions occur in periodic cycles would be premature, to say the least: Four points are hardly enough to prove periodic behavior. Meanwhile, the data on crater ages are too sparse to show any real peaks, save those of the past few million years and the three craters with ages of about 38 million years. No wonder, then, a skeptic would say, that the individual crater ages can be interpreted as showing periods of either 28.4 million or 31 to 32 million years!

To this skepticism, those deeply involved in the argument would reply that the data as now known are highly suggestive of periodic behavior, and that in view of this suggestiveness it is worthwhile to consider the theories that are based on the hypothesized cyclical recurrence of mass extinctions and impact craters. They are, after all, engaged in scientific inquiry, and need not demand complete proof before proceeding to see where speculation will lead them. All that then remains is for these scientists to examine the theories, and to see whether they can find support in areas beyond the controversial analyses that we have just examined.

3

WHAT CAUSES MASS EXTINCTIONS?

THE REALIZATION BY paleogeologists that mass extinctions have occurred many times on Earth within the past few hundred million years, coupled with the discovery of the excess iridum within the thin layer of clay that marks the Cretaceous-Tertiary boundary, has spawned many theories to explain mass extinctions through impact. These theories start with the hypothesis that a single impact produced an iridium-rich layer 65 million years ago, and then extrapolate to other impact events, seeking a single cause for all known mass extinctions. According to these theories, objects of significant size strike the Earth at regular, though widely spaced, intervals of time. These impacts create effects that extinguish great numbers of species of life on Earth.

To analyze these theories, we must concentrate on a few obvious questions that they raise: *What objects* collide with Earth? *What effects* do they produce? *What causes* the objects to strike the earth? Do these causes suggest *periodic recurrence?*

Asteroids and Comets

The objects of significant size most likely to collide with Earth, over time spans measured in millions of years, are the asteroids and comets. Tiny asteroidlike objects called meteoroids enter the earth's atmosphere by the billions every day.

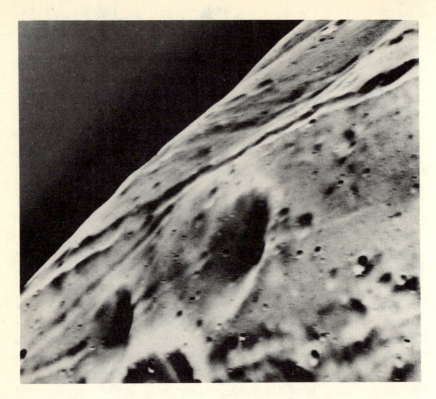

Figure 6—Phobos, the innermost of Mars's two small satellites, is thought to be an asteroid that was captured into orbit around Mars. We have yet to send a spacecraft close to any asteriod, but this photograph of Phobos can serve an our prototype. Note the numerous small craters and long grooves, testimony to impacts from smaller objects on Phobos after it had formed. (NASA photograph.)

Frictional heating caused by the meteoroids' passage through the atmosphere produces visible meteors, or "shooting stars." Some large meteoroids may reach the earth's surface without burning up completely in the atmosphere, leaving meteorites, which we may subsequently discover. Both the asteroids and comets have a long history in astronomical lore, but both have only recently begun to reveal their true nature to modern astronomical research.

Asteroids, whose name means "like a star" in Greek, resemble stars only in their pointlike appearance in a telescope; they are members of the solar system, orbiting the sun, controlled by the sun's force of gravitation and by their momentum in orbit. The largest asteroid, Ceres, has a diameter of 1,000 kilometers, nearly one-third the diameter of the moon; the next largest, Pallas and Vesta, have diameters of 600 kilometers (about as large as Wyoming). Typical meteoroids, which are basically mini-asteroids, range in size from a few meters down to a fraction of a millimeter in diameter.

Ceres, Pallas, and Vesta all orbit the sun between the orbits of Mars and Jupiter. The same holds true for most of the thousands of known asteroids, whose diameters can be as small as one kilometer (the size below which we cannot detect asteroids, which can be seen only because of the sunlight they reflect). Millions of asteroids too small to spot are presumed to be members of the "asteroid belt," the swarm that orbits the sun (including all of the largest asteroids) beyond Mars, at 2 to 3.5 times the Earth–sun distance.

Even Ceres is too small to be seen in our largest telescopes as anything but a point of reflected sunlight. To find out more about asteroids, we shall have to send spacecraft close to them. Such a trip is on the agenda of NASA's Project Galileo mission to Jupiter, which, with a small course correction, should pass close to the asteroid Amphitrite in 1988, sending us images and spectroscopic observations—color-by-color analysis of the reflected sunlight—before embarking on its main duties around the giant planet. Amphitrite, about a hundred kilometers in diameter, happens to occupy a point in its orbit almost in the path of the Galileo spacecraft, so for a relatively small cost, we can hope to find out much more about the composition of these long-known but still-mysterious celestial objects.

Apollo Asteroids and Amor Asteroids

A group of rare asteroids, called the Apollo asteroids after its best-known member, have orbits that intersect with that of Earth. The earth has a slightly elliptical (oval) orbit around the sun, and varies its distance from 1.02 times (in July) to 0.98 times (in January) the average Earth–sun distance, which astronomers call one astronomical unit, or A.U., equal to 149,597,900 kilometers, or about 92.9 million miles. The Apollo asteroids have elongated orbits, approaching the sun to within less than one A.U. and receding to distances of 3 to 4 A.U. About thirty Apollo asteroids have been discovered; they range in size from Hephaistos, which is 9 kilometers across, down to objects only half a kilometer, or even less, in diameter. One of the Apollo asteroids, Icarus, 1.4 kilometers in diameter, has an orbit that carries it far inside the orbit of Mercury, to within 0.2 A.U. of the sun, and well beyond the orbit of Mars, to a maximum distance of 2 A.U.

In addition to the Apollo asteroids, which actually cross Earth's orbit, a related group of asteroids, the Amor asteroids, have orbits that nearly, but not quite, intersect with Earth's. The largest of the Amor asteroids, Eros, 15 kilometers across, has an orbit that carries it from 1.78 A.U. to just 1.13 A.U. from the sun, so Eros can approach the Earth to within 20 million kilometers—a close encounter by solar-system standards, though still fifty times the distance from Earth to the moon. Icarus, whose orbit *does* cross Earth's, approached our planet to within a million kilometers in 1930, and has been closer than that in ages long past. This is still a long way from direct impact, but there is lots of time to wait in the solar system.

Eugene Shoemaker estimates that more than a thousand Apollo asteroids exist, with most of them much smaller than 100 meters in diameter. From the viewpoint of a theory that attempts to explain mass extinctions through extraterrestrial impacts, it is the large, already-known Apollo asteroids—the asteroids that cross Earth's orbit—that deserve the greatest attention. However, Amor asteroids can *become* Apollo asteroids as their orbits are perturbed by the planets' gravitational forces, and indeed both the Apollo and Amor asteroids were probably "ordinary" asteroids until they were perturbed from larger into smaller orbits.

Comets

In contrast to asteroids, *comets* typically have orbits that carry them much farther from the sun than the Earth–sun distance. The trillions of comets orbiting the sun at immense distances—tens of thousands of A.U.—collectively form the Oort cloud, named after the Dutch astronomer Jan Oort, who first proposed the existence of such a huge swarm of comets in the outer solar system.

Most comets move around the sun far outside the orbits of all the planets; the exceptions consist of comets that have an average distance from the sun about equal to the distances of Jupiter and Saturn, 5.2 and 9.5 A.U. respectively. Within this exceptional group, a few comets have elongated orbits that carry them near Earth's orbit, or even inside it, at their points of closest approach to the sun. The best-known of all comets, Halley's Comet, belongs to this limited category. Halley's Comet takes 76 years to orbit the sun, and spends most of this time outside the orbit of the planet Uranus. But at its point of closest approach, Halley's Comet passes inside Earth's orbit, and indeed inside the orbit of Venus, gaining in velocity as it approaches the sun, until it swings around the sun and heads outward again, vanishing into cold obscurity before returning to amaze a later generation of humanity.

Comets and asteroids have important common features, and equally important dissimilarities. Although the orbits of a few comets and a few asteroids are rather similar, most asteroids orbit the sun between the paths of Mars and Jupiter, and their orbits have significant, but not enormous, elongations.

Most comets, however, remain far beyond all the planets'
orbits, and they have much more elongated orbits than planets
and asteroids do. These greatly elongated orbits reflect the way
in which the comets formed from material orbiting the sun-to-
be. As planets agglomerated from smaller, cometlike objects,
their orbits became more nearly circular.

With the few exceptions that we have noted, all comets
remain at distances of tens of thousands of A.U. from the sun.
If we picture the solar system as a grapefruit surrounded by
darting insects, then Earth would be a tiny fly, orbiting the
grapefruit at a distance of five meters. Jupiter would be a fairly
large mosquito, 25 meters from the sun. The asteroids would
be nearly invisible mites at distances of four to twenty meters,
and the comets, even smaller, would be at distances of many
kilometers! These immense distances are what make any collisions between solar-system objects unlikely, and any theory
that depends on such impacts must take this into account. In
seeking to explain the record of mass extinctions, and of impact
craters on Earth, these theories look for a reason that some of
the tiny mites would dart into the inner solar system, where a
nearly infinitesimal portion of them might collide with our
planet.

What Are Comets and Asteroids Made of?

The characteristics of their orbits do not provide the best
way to distinguish between comets and asteroids, because
their orbits can change with the passage of time. More important differences—and resemblances—are found in the *composition* of these objects, though the composition of comets may
change over time as their outer parts wear away. This composition determines the density of matter within the object, and
therefore the energy released in a collision with an object of a
given size and velocity (such as the earth).

Comets, whose name derives from a Greek word meaning "long-haired," are the much-feared interlopers of astronomical history, known for their strange, "hairy" appearance. Comets consist of volatiles, substances that evaporate readily, frozen around dust and rock particles. In a comet, the volatiles are mostly methane, ammonia, water, and other compounds of hydrogen with carbon, nitrogen, and oxygen atoms.

If a comet passes relatively close to the sun—say, within the orbit of Jupiter, 5.2 A.U. from the sun—the sun's heat releases some of the volatiles in the form of gases. The newly vaporized gas hovers around the cometary nucleus—the frozen part of the comet. This gas forms a gauzy coma (from the Greek word for "hair") surrounding the nucleus, and some of it reaches great distances from the comet, producing the comet's tail. Sunlight reflects from the coma as well as from the tail, which consists of gas so diffuse that it would qualify as a vacuum on Earth. The image seen from Earth, because of its fuzzy or hairy appearance, differs from that of any star or starlike object, planet, or asteroid. Comets have been seen throughout history as disruptions of the fabric of heaven, portents of doom— though just whose had to wait until after the fact to be proven! Since comets only rarely come close to the sun and Earth, each cometary appearance was a frightening, unexpected event, until Edmund Halley correctly worked out the orbit of the comet now bearing his name, and predicted the date of its following apparition.

Asteroids, unlike comets, do not have volatiles that can escape when heated. Because asteroids move in orbits that *always* lie close enough to the sun to allow the coma and tail of a comet to develop, it is reasonable to suppose that asteroids may once have had volatiles but have lost them through continued exposure to the sun's heat. Astronomers would agree with this general statement but would debate whether the asteroids *first* formed and *then* lost their volatiles, or

Figure 7—Comet Mrkos, a relatively small comet no more than a few kilometers in diameter, appeared in 1957. As it neared the sun, the comet released gas and dust, which formed a fuzzy coma around its basic nucleus, as well as long tails of gass and dust that streamed behind for millions of kilometers. This photograph shows both the gas tail, indentifiable by its complex ripples, and the dust tail, which simply reflects light diffusely. The dust tail appears just below the tail (actually several tails) of gas. (Palomar Observatory photograph.)

whether their material had lost its volatiles even before the asteroids formed. The latter view is now held by the majority: It pictures matter swirling in orbits around the sun-in-formation, some 4.6 billion years ago. As the sun's heat warmed the inner parts of the solar system, volatiles escaped from these regions simply because the matter there turned from solid to gas and then evaporated. Hence the inner solar system—the parts inside the orbit of Jupiter—became volatile-poor, while objects in regions farther from the sun, where the giant planets Jupiter, Saturn, Uranus, and Neptune formed, retained their volatiles.

The Formation of the Solar System

The most widely held current views of how the solar system formed imply a basic distinction between the inner solar system and the volume farther from the sun, a distinction based on the greater solar heating of the inner system. The planets Mercury, Venus, Earth, and Mars—as well as the moons of Earth and Mars, and the asteroids that orbit between Mars and Jupiter, and even closer to the sun—all have lost their volatile component. When we look at the four giant planets, which have retained their volatiles, we see that this is more than a "component." Compounds of hydrogen, mostly methane and ammonia, form 90 percent of Jupiter's mass, while pure helium forms 9 percent, leaving 1 percent for nonvolatiles. If Earth had retained its "fair share" of volatiles, rather than losing them because of the sun's heat as the planet formed, it would have a hundred times its present mass and might resemble a giant snowball of frozen methane and ammonia, with some "impurities" that we call the rocks and metals of Planet Earth.

This description of the might-have-been Earth is in fact a description of the comets, except that each comet has a diameter of only a few, or perhaps a few dozen, kilometers. We may picture a cometary nucleus as a primordial iceberg, formed along with the solar system and fundamentally unmelted and unchanged through the intervening 4.6 billion years.

When we consider comets as a group, one startling fact may seem temporarily overwhelming: their number. It is the huge *number* of comets—trillions upon trillions of them in the sun's Oort cloud—that promotes the hypothesis that mass extinctions on Earth arise from collisions with comets, perturbed from their usual orbits by gravitational forces from an interloping object.

Where Did the Comets Form?

Picture the solar system in formation as a swirling disk of gas and dust; a diffuse, rotating pancake with a diameter twenty thousand times the Earth–sun distance (20,000 A.U.), spinning more rapidly at its center, where most if its mass is concentrated, than at its outer edges. A few million years previously, within a much larger interstellar cloud of matter, this "dusty pancake" had begun to contract as a clump of gas and dust because of its self-gravitation—the pull of each piece of matter in the clump on the other parts. The contraction made the clump spin more rapidly, and led to the flattening of the contracting clump, because the clump's "centrifugal force" tended to support matter in directions perpendicular to its axis of spin.

A few million years more saw the pancake shrink in size, bringing most if its material much closer to the central condensation and thereby increasing the density of gas and dust within the inner parts of the rotating clump. Once the clump had contracted to a diameter of about a hundred A.U., the density rose to the point where particles of gas and dust began to collide with significant frequency. Such collisions among the particles orbiting the protosun (the sun-to-be) produced accretion, the sticking together of atoms, molecules, and still larger dust particles. From this accretion grew snowball-size and even larger objects, which, in this theoretical model of how the solar system formed, bear the name *cometary planetesimals*. *Planetesimal* means "tiny planet," but we are in fact describing the "seeds" of planets, the small clumps, each a few kilometers across, that collided to produce larger objects in orbit around the sun.

This process, it is now believed by most astronomers, involved not millions or billions, but many *trillions* of objects a few kilometers in diameter, all orbiting the sun at distances from a few tenths to a few hundred times Earth's present distance. But a key factor affecting the closer planetesimals more than the distant ones was the greater and greater amount of heat being produced by the sun. This heat arose first from the contraction of the protosun, and then, when the sun's center had grown so hot that nuclear fusion began, from the release of kinetic energy by nuclear-fusion reactions.

So the sun's heat warmed the inner regions of the solar system most, the middle regions less, and the outermost regions least of all—a situation that persists to this day. Within about 4 A.U. of the sun, any cometary planetesimals, and any planets that formed from them, began to lose their volatiles as the sun warmed them. The methane, ammonia, and water present within the inner solar system evaporated as the sun began to shine, turning from solids into gases that escaped forever, along with helium and neon gas, from the objects that could retain these volatiles only so long as they remained frozen. Today Earth and the inner planets retain only a tiny fraction of their original share of volatiles, which they trapped beneath their surface layers and later spewed out in volcanoes to form their atmospheres and, in Earth's case, seas of water.

Farther out from the sun, the volatiles were retained, so that the giant planets Jupiter, Saturn, Uranus, and Neptune consist largely of ammonia, methane, and water. These planets are giant spheres of hydrogen compounds mixed with helium and neon. If we were to descend into Jupiter, the planet's gaseous outer layers would thicken into a denser and denser muck until we reached the planet's core, hydrogen frozen around an iron-rock center reminiscent of Earth. More than 90 percent of Jupiter's 318 Earth-masses of material (including nearly 99 percent of its outer, gaseous layers) consists of hydrogen and helium volatiles. If we heated Jupiter to the point that these volatiles could escape, as they did from the earth 4.6 billion years ago, we would have an Earthlike planet with perhaps 10 times the Earth's mass. Jupiter is Jupiter because it is 5.2 times Earth's distance from the sun; Earth owes its volatile-poor character to its greater exposure to solar heat.

The dividing line between volatile-rich and volatile-poor apparently lies just outside what we call the asteroid belt, the region between the orbits of Mars and Jupiter, at distances from 2.2 to 3.8 A.U., that contains most of the known asteroids. These asteroids are at an average distance from the sun of 2.8 A.U. and —like Mercury, Venus, Earth, and Mars—have lost most of their original volatile material. Their average density resembles that of the rocks near Earth's surface, about three times the density of water. In contrast, the nuclei of comets, like the giant planets, consist mainly of volatiles and frozen water, methane, and ammonia.

Pluto, the outermost planet, likewise consists of frozen volatiles; its density, only seven-tenths the density of water, shows its resemblance to a frozen iceberg, but one less dense than a terrestrial ice floe (0.92 times the density of water). Pluto seems to be composed of materials just like those in comets, and might be considered the largest comet of them all—about the size of our moon. Other objects suspected to be former comets are Phoebe, Saturn's outermost satellite; and Charon, usually classified as an asteroid, which moves in one of the largest asteroidal orbits, almost the size of Saturn's orbit.

Cometary Orbits Around the Sun

The orbits of all the planets save Mercury, the innermost, and Pluto, the outermost, are ellipses, but with so little elongation that an untrained eye cannot distinguish them from circles. Mercury and Pluto have modestly elongated elliptical orbits, whereas even the largest asteroids have *significantly* elongated orbits, often much more so than those of any planet. Such orbits may well have been the rule for all cometary planetesimals; the process of combining into larger and larger objects apparently also reduced the eccentricity of the larger objects' orbits, so that today the seven large planets have nearly circular orbits. Cometary orbits show a different pattern, the result of a different history since the comets formed.

Astronomers once thought that the swarm of solar-system comets in the Oort cloud, moving in orbits tens of thousands of A.U. from the sun, formed at these immense distances within the contracting, rotating pancake of matter—a tiny part of an immense galaxy—that became our solar system. However, calculations strongly indicate that the density within this solar nebula fell too low for anything as large as a comet—1 to 100 kilometers in diameter—to form from the sticking together of particles that collided; the density was far too low for such accretion to occur efficiently. Modern theories of cometary formation therefore assign their origin to the inner parts of the solar system—five to fifty times Earth's distance from the sun—where the giant planets and Pluto were created.

The region of the solar system that is 5 to 50 A.U. from the sun never grew so warm that the vaporization of volatiles occurred, but the relatively small volume of this region allowed significant accretion of matter. By the trillions upon trillions, cometary planetesimals grew from the accretion of volatile-rich material in this volume. Most of these objects collided to form the giant planets, which contain many times more mass than remains in cometary nuclei.

Once the giant planets had grown to roughly their present sizes and masses, their strong gravitational forces produced an important effect on the cometary orbits. One by one, sooner or later, the cometary planetesimals passed close to one of the giant planets. Some of these close encounters drew the comets into the planets, adding to their mass, but those that did not flung the comets into much larger orbits by a "gravitational slingshot" effect: The comets accelerated because of the planet's gravitational force, but passed *around* the planet and out into space at a higher velocity than they would otherwise have had.

Within a few million years after the giant planets had formed, they had swept their part of the solar system almost clean: Several trillion cometary nuclei had been forced from their birthplace into near-perpetual exile in the frozen outlands of the sun's gravitational dominion, while others were banished from the solar system altogether. The typical encounter with a giant planet gave the comet an average distance from the sun of 10,000 to 20,000 A.U., but with an orbit so elongated that the comet would approach the sun, once every few million years, to one-hundredth of this distance. Even this point of closest approach, a few hundred A.U., left the comet far outside the orbits of all the planets (Pluto orbits at 40 A.U.), incapable of colliding with any of them, and never in danger of losing any of its volatile-rich material. In our search for an explanation of cometary impacts, we must ask, what could happen that would give a comet a chance to collide with Earth?

The Later History of Comets

Were the sun alone in the Milky Way, we would know comets only as left-over representatives of the pieces from which the sun and its planets were formed. We would have no lore of comets as harbingers of disaster, and Edmund Halley would only be (justly) famous for his efforts at mapping the heavens, charting the seas of Earth, analyzing the motions of the moon and the planets, measuring changes in the earth's magnetism, and helping Sir Isaac Newton to formulate his theory of gravitation. But the stars in the Milky Way galaxy, each orbiting the galactic center, do not have identical orbits. As a result, they move independently of one another as each follows its own majestic path around the Milky Way. Some stars occasionally pass relatively close to the sun, to within a few hundred thousand A.U. of our star. This effect, which the astronomer Patrick Thaddeus, of Columbia University, calls "the pitter-patter of passing stars," creates the chance to divert comets into new orbits.

As stars move past the sun, at thousands of times the planets' distances, the changing amount of gravitational force that they exert on comets will affect cometary orbits, pulling some comets into new orbits that carry them into the realm of the planets. Jack Hills, of the Los Alamos Scientific Laboratory, has calculated the strength of this effect. In 1981, Hills introduced the concept of a comet shower, a pulse of millions of comets diverted into the inner solar system by a passing star. Here the comets are likely to be reaffected by the gravitational force from one of the giant planets—in particular, from Jupiter, the most massive. As a result, although the comets will continue to orbit the sun, their orbits will keep the comets at average distances from the sun like those of the giant planets' orbits, 5 to 30 A.U. Like the giant planets, the comets will take from a few to a few dozen years to orbit the sun once. However, the comets' orbits will typically be more elongated than the planets' because of their recent perturbations by gravitational forces. Comet Borrelly, for example, orbits the sun once every seven years, with an average distance from the sun of 3.7 A.U., but it goes as far away as 5.9 A.U. and approaches the sun to within a mere 1.45 A.U.

Hence Comet Borrelly cannot last for long, in astronomical terms. Each close passage to the sun releases some of its frozen volatiles, so the comet is doomed to vanish—in a thousand, ten thousand, or a million years, depending on how tightly frozen its nucleus is. The comet will either evaporate completely or will split into several large pieces, which themselves will eventually undergo the vaporization. The same holds true for all the short-period comets, those diverted by passing stars into close, planet-distance orbits. If the comet contains an inner core of rock, that core may survive even after the volatiles have evaporated. In other words, some comets may eventually become asteroids. Comet Encke, which moves in a highly elliptical orbit that takes it from 4.1 A.U. to within 0.34 A.U. of the sun (inside Mercury's orbit) is believed to be making this transition even now.

Although the *material* in the short-period comets will eventually vaporize, their *orbits* are fairly stable: They are unlikely to change so much that they could collide with Earth, even on astronomical time scales. For those comets whose orbits pass inside Earth's, the lack of collision is a matter of luck. The volume of the inner solar system, though tiny in comparison with the volume that includes the sun's entire family of comets, is still enormous, so large that the few dozen short-period comets that do have orbits within this volume never strike Earth even in the course of millions of orbits around the sun. If we consider not a few dozen but a few billion comets, however, the odds will shift, and the collision of one or more comets with Earth becomes likely. Here we have the root of the astronomical explanations for the mass extinctions of life on Earth.

Cometary Collisions With the Earth

What would be the effect of a comet striking the earth? And how likely would such a collision be, if a huge number of comets were diverted into short-period orbits? We can now answer these two questions more readily than we can the key question that underlies comet-impact theories: What could move a billion comets into short-period orbits? The answer to the first two questions may, however, help us to understand the debate about the third.

To take the question of probabilities: Space is mostly empty, but even so, the planet Earth fills a finite volume within a much larger system of matter. The Earth–sun distance, 149.5 million kilometers, equals 23,400 times Earth's radius (6,378 kilometers). We can therefore calculate, using high-school math, that Earth occupies 1 part in 13 trillion of the volume of a sphere centered on the sun and with a radius equal to the Earth–sun distance. This calculation may make a collision between Earth and a comet on a random trajectory through the inner volume of the solar system seem extremely unlikely, but we should not forget that our planet's gravitational force can attract the comet.

If we picture a sphere centered on Earth, but with a radius 60 times the planet's radius (equal to the moon's distance from Earth), this sphere has a volume that is the cube of 60—or 216,000—times larger than Earth. So the probability that a comet moving at random through the inner part of the solar system will pass through this sphere is not 1 in 13 trillion, but 216,000 times greater: 1 in 60 million. And indeed, if a comet *does* have an orbit that brings it about as close as the moon, the probability that it will be diverted by our planet's gravitation to collide with it is high. If the distance of closest approach, not counting the effect of Earth's gravity, were much greater than this, then the planet's force of gravitation probably would not be strong enough to attract the comet into an Earth-intersecting trajectory. More exact estimates—but they are still estimates—rate the chance of a collision of a comet with Earth as about 1 in 200 million for each time that the comet passes inside Earth's distance from the sun.

Perturbed Cometary Orbits

Comets that reach the planetary region of the solar system feel a strong perturbing effect from the planets' gravitational forces as they round the sun and head outward. Jupiter, having 318 times the mass of the Earth, has by far the most powerful effect on the comets; Saturn, with 95 Earth-masses, ranks second; the other two giant planets, Uranus and Neptune (15 and 17 Earth-masses, respectively) come next; while Earth and the other less massive planets have almost no effect at all on cometary orbits. Each time a comet enters the planetary region, the gravitational forces from Jupiter and Saturn are likely to perturb it into an orbit that keeps the comet far beyond all the planets, just as happened when the comets formed, 4.6 billion years ago. Therefore most comets entering the planetary region have only one orbital passage in which a collision can occur. A tiny fraction of these comets will be captured into short-period orbits, and a small fraction of these may follow a path that intersects repeatedly with Earth's orbit, thereby giving such rare comets many chances at collision with our planet.

Therefore, if we imagine a perturbation that suddenly (in astronomical terms) sends a billion comets into the planetary region, the result would be a "pulse" of possible colliders just as soon after the perturbation as it takes the comets to arrive—about a million years for comets perturbed at a distance of 20,000 A.U. from the sun. This pulse would decline sharply as most of the comets were again expelled, but a few would acquire short-period orbits, remaining in Earth's vicinity for hundreds of thousands, or even a few million, years. Throughout this period the number of those comets, some of them capable of acquiring Earth-crossing orbits, also would "tail" off. This "tail" provides a way to reconcile the record of mass extinctions, which does not show any perfect periodicity, with a theory that cometary orbits are perturbed at intervals of 26 to 33 million years. The perturbations may occur with complete regularity, but their effects—impacts on Earth—would have a time delay that spreads, and by a varying amount, over as much as one to two million years.

Collisions With Asteroids

We can also analyze the probability of a collision between Earth and an asteroid whose orbit passes *inside* Earth's orbit. These orbits typically have periods of 1 to 3 years, averaging about 1.8. Two crossings of Earth's orbit occur during each orbit of such asteroids and each time the asteroid crosses Earth's orbit, the odds are 1 in 300 million that a collision will occur. From this we can calculate that each such Earth-crossing asteroid will collide with our planet after about 270 million years.

Recall that the largest Apollo (Earth-crossing) asteroids, Sisyphus and Hephaistos, have diameters close to 10 kilometers; and the largest Amor (nearly Earth-crossing) asteroids, Ganymed and Eros, have diameters of 40 and 20 kilometers respectively. Therefore, if the present situation represents a typical time in the history of the solar system, we can expect an impact with a 10-kilometer or larger asteroid once every few hundred million years. This could explain some, but not most, of the impacts found in a 28- or 32-million-year cycle. If we look for objects that could produce impacts not every few hundred million but every few *tens* of millions of years, we must look to the comets.

Additional Evidence of Impact: The Microtektites

The iridium-rich layer from the Cretaceous-Tertiary boundary provides the best evidence for the impact of a large object with Earth. We do have additional evidence for such impacts, including one at the time of the Eocene-Oligocene transition, 38 million years ago.

This evidence consists of glassy objects called tektites, whose name means "molten" in Greek. Geologists now generally agree that tektites have arisen from the melting of surface material on the earth during impact events. Tektites appear in only four areas on the earth's surface, called tektite strewn-fields. These four strewnfields, dated by radioactive-decay techniques, have ages of 0.7, 1, 14.7, and 37 million years. The latter two ages correspond fairly well to the two most recent peaks in the mass-extinction data, and the last of these four ages lies within a million years of the date of the Eocene-Oligocene boundary.

Microtektites—tektite spherules only a few thousandths of a centimeter in diameter—have been found in many ocean cores drilled in the strewnfields, and have been studied by an expert named (appropriately) Billy Glass, of the University of Delaware. The estimated total of tektite material in the largest and youngest strewnfield, thought to be less than a million years old, equals 100 billion kilograms, about one four-thousandth of the mass of the object hypothesized to have impacted at the end of the Cretaceous period. However, a high-velocity object can affect a mass of material on Earth far greater than the mass of the object itself, so that an object much less massive than 100 billion kilograms could indeed melt 100 billion kilograms of terrestrial material. If we assigned a mass of ten billion kilograms to the impacting object, we would be led to imagine a meteoroid or asteroid with a diameter of only a few hundred meters, still capable of melting a large amount of matter and splashing it over a sizable fraction of the earth's surface.

Tektites therefore confirm the notion that extraterrestrial objects do strike the earth from time to time, and the coincidence in age between the oldest of the four strewnfields and the mass extinction 38 billion years ago supports the hypothesis that an impact occurred near the time of the Eocene-Oligocene extinctions. But the evidence from tektites hardly proves that large impacts have occurred in a periodic manner. We must rate tektites as intriguing records of impact, but not as confirmation of the impact hypotheses described in this book.

Asteroidal and Cometary Impacts

In attempting to investigate the hypothesis of periodic impacts, we must make a quick calculation that compares the effects of asteroidal and cometary impacts to see whether comets, with lesser masses than asteroids, can produce equal results.

The effect of any impact on the earth's surface will increase in proportion to the *mass* of the impacting object and to the *square of its velocity* with respect to the motion of Earth. The object's kinetic energy (the energy of its motion) depends on this product, and it is the energy brought to Earth, no matter how mass and velocity combine, that does the damage.

Asteroids and comets that could strike the Earth have velocities (with respect to our moving planet) that vary from only a few up to about 60 kilometers per second, depending on whether they approach via head-on, overtaking, or intermediate paths. We know this because comets and asteroids have roughly the same speed in orbit as Earth does (30 kilometers per second) but their velocities can add to (in head-on collisions) or subtract from (in overtaking collisions) the planet's orbital velocity. The average velocity with respect to Earth will be that of an intermediate collision, 30 kilometers per second.

Because asteroids have been affected in their orbits by the gravitational pull of the giant planets, we can think of them as "socialized" into the sun's family; almost all of them orbit the sun in the same direction that the planets do, whether or not they had this orbital direction originally. Asteroids are therefore unlikely to engage in head-on collisions with Earth. Comets, by contrast, move in orbits that have never been "socialized" since they first formed, billions of years ago. Their orbits have random directions with respect to Earth's direction of motion, so a head-on cometary collision is as likely to occur as an overtaking one. An asteroid collision, most likely to be an overtaking situation, will occur with an average relative velocity of 10 to 15 kilometers per second. In contrast, a comet's average relative velocity in a collision will be close to 30 kilometers per second. Therefore a comet with the same mass as an asteroid could, on the average, produce four to nine times as severe an effect. Put another way, a cometary collision and an asteroid collision have the same average effect if the comet has only one-fourth to one-ninth the asteroid's mass.

The mass of a colliding object varies in proportion to the cube of the object's radius times its density of matter. Cometary material, like that of Pluto, is thought to have a density of about 0.7 times the density of water, about one-quarter to one-fifth of the density of the rocky asteroids. Inserting these numbers, we find that the greater average velocity of a comet, relative to Earth, nicely compensates for the lower density of cometary material, so to a first approximation we can say that, on the average, a comet and an asteroid of the same *size* will have the same overall effect when they collide with Earth, even though the asteroid would have four to five times the comet's mass. We may note in passing, however, that in view of the lower density of comets, it would take a 12-kilometer comet to bring to Earth the same amount of iridium that a 7-kilometer asteroid would, assuming the fraction of iridium is the same in both objects.

Effects of Impact on the Earth

Picture an onrushing comet or asteroid that approaches Earth at a relative velocity between 10 and 50 kilometers per second, so that the object arrives from the moon's distance in a matter of a few hours—not much of a warning even if anything were to be done about it. The comet or asteroid would part the atmosphere like a superprojectile, leaving a hole through the air as wide as itself. Even though air rushes back into the hole at the speed of sound, it would take tens of seconds for the hole to disappear. Upon striking the earth, the object would decelerate to zero velocity only after depositing most of its kinetic energy in the material that it encounters. Thus, the object would stop only after it has encountered a total mass of material several times its own mass—a bit more than this if we describe a high-velocity comet, a bit less if we consider a relatively low-velocity asteroid impact.

The oceans have an average depth of 7 kilometers, so the ultimate result would be much the same whether the object struck land or water; if the latter, it would plow a hole through the water—after ripping through the atmosphere—and then excavate a crater several kilometers deep and 50 to 100 kilometers wide in the earth's crust. The water around the comet would be vaporized at once, doubling the water-vapor content of the atmosphere. Applying our knowledge of earthquakes to the object's kinetic energy, we find that the earthquake from such an impact would release 100 billion times more energy than did the 1906 San Francisco earthquake. Another effect of an impact in the oceans would be a tsunami (popularly misnamed a "tidal wave") about a kilometer high, which would roll across the seas at 1,000 kilometers per hour and cause widespread destruction in areas within a hundred kilometers or so of the coast.

Those are but the side effects; most of the damage to life on Earth would arise from dust. A 10-kilometer object would move about 200 cubic kilometers of rock, more than a thousand times the amount excavated over ten years to make the Panama Canal. The impact would heat this material instantaneously, spraying it sideways and upward from its crater. Matter heading upward would encounter no resistance from either ocean (if the comet struck there) or atmosphere, for they would have been thrust aside sufficiently for the material to rise unimpeded. The pulverized particles would each acquire a ballistic trajectory, like the path of a rocket once it has left the atmosphere and its engines have shut down. Each particle rising high above the earth would orbit our planet. Some of the particles would fly off into interplanetary space, but most of them, still held by Earth's gravity, would fall back onto the top of the atmosphere, at a point far from the hole through which they emerged. The heavier particles would fall through the atmosphere at points all around the globe, but the lighter ones would float like oil on water, suspended in the stratosphere as fine-grained, low-mass dust grains. The result would be death for many species of life on Earth.

When scientists study the nuclear winter that would arise from a large-scale exchange of nuclear weapons, they must examine the effects of radioactive debris, of ultraviolet radiation that reaches the surface because of the destruction of the ozone layer, of gamma radiation from the nuclear weapons, and of acid rain from nitrous oxide. Hazardous as all these effects would prove to be, they pale in significance beside the chief effect produced both by thermonuclear bombs and by a large extraterrestrial impact—the blotting out of the sun.

Cometary Impacts and Nuclear Winter: Three Months of Darkness

Thus a long period of darkness would result from the dust lifted 20 to 40 kilometers into the atmosphere, above the tropopause, usually 10 kilometers high, that puts a "lid" on all weather patterns. Any dust that enters the stratosphere (above the tropopause) takes months to settle to the surface, because it cannot be brought down by rain, since water vapor does not rise that high. In a nuclear exchange, most of this dust would arise from the fires started by nuclear weapons, but in the extraterrestrial-impact scenario, the dust would arise, even more straightforwardly, from the vaporization of many cubic kilometers of matter at the site of the impact.

Five scientists who have made calculations of this effect— Richard Turco, O. Brian Toon, Thomas Ackerman, James Pollack, and Carl Sagan (referred to by the acronym of their last names as TTAPS)—have considered what would happen if about a billion tons of dust and smoke were injected into the stratosphere. (A 10-kilometer asteroid or comet would produce ten to a hundred times *more* dust than this.) Actually, the sooty, carbon-rich particles from fires can absorb sunlight more efficiently, and remain suspended longer in the upper atmosphere, than the quartzlike silicon-oxygen dust particles from a cometary or asteroid impact. Thus the smoke and dust from a nuclear exchange could still have a devastating effect on planetary life.

TTAPS sum up the effects as "significant surface darkening over many weeks, subfreezing land temperatures persisting for up to several months, large perturbations in global circulation patterns, and dramatic changes in local weather and precipitation rates." As long as large numbers of dust particles remained in the upper atmosphere, they would absorb all sunlight, heating the upper atmospheric layers but cooling layers near the surface, which would no longer receive the sun's light and heat.

The dust particles would eventually rain out of the upper atmosphere, however, as they collided with each other and adhered to form larger particles, which cannot remain suspended in the rarefied air, and would sink rapidly toward the surface. Calculations made by the TTAPS group, with the collaboration of Christopher McKay of the University of Colorado, and Michael Liu of Informatics, Inc., in Palo Alto, show that the rate at which sunlight returns to the surface depends only slightly on the density of dust particles in the upper atmosphere. If the density of dust is relatively low, it blocks less sunlight, but it takes the dust longer to collide, stick together, and settle to the surface. If the density is high, collisions occur more often and each dust particle has a higher probability of falling to the surface by sticking to an increasingly heavy bit of floating material. Hence if a vast quantity of dust were to enter the upper atmosphere and spread around the globe, it would remain there for "only" three to six months.

These months would be unpleasant for life as we know it. For a number of weeks, the intensity of sunlight on the Earth's surface would fall to below 1 percent of its normal value, low enough to halt photosynthesis among the phytoplankton in the oceans. David Milne and Christopher McKay, of the University of Colorado, have calculated that somewhere between 10 and 100 days without sunlight would be enough to kill all the phytoplankton. Since these tiny plants form the bottom of the entire food chain in the seas, their loss would in turn cause the disappearance of all smaller marine animals (zooplankton) unable to survive for a few weeks without food. It is precisely such plankton whose disappearance is so noticeable at the Cretaceous-Tertiary boundary, where the iridium layer that led to the impact theory was found.

The interval of 10 to 100 days will occur within the interval of 3 to 6 months predicted for the strong attenuation of sunlight by dust in the atmosphere. The other significant effect this dust produces, the cooling of the earth's entire surface, also has pernicious effects on life. The TTAPS group estimated that the average surface temperature would decline, within a few days after the dust entered the atmosphere, by some 40 degrees Celsius (70 degrees Fahrenheit): from 10° C (50° F) to −30° C (−22° F). This effect would be most pronounced over land masses, because the oceans store heat so well that they would cool by only a few degrees, and they moderate the cooling of coastal areas. Even so, storms over the oceans would create adverse conditions within at least the upper levels of the seas. But the label *nuclear winter* really applies only to inland areas; for the bulk of the world, the oceans and coastal regions, a better description would be the "three months of darkness."

From the Darkness: Life!

From such darkness, few of us would expect to emerge alive. The same holds true for most species of plants and animals even before the changes in weather patterns, adding to the new burdens of life, are taken into account. The storms caused by sudden changes in temperature would be most pronounced at the coastal margins, where the newly frigid continents met the still-warm oceans. In contrast, the effects of increased ultraviolet radiation would be felt everywhere except in the ocean depths. An impact that raised dust to block visible sunlight would also produce molecules that combine with the ozone in the Earth's stratosphere, removing the ozone molecules and, with them, our shield against solar ultraviolet. Thus in deadly irony, as the dust settled and allowed visible sunlight once again to reach the surface, the lack of an ozone layer would permit the penetration of short-wavelength ultraviolet radiation. Such radiation would harm the immune systems of all mammals, and probably of other animals as well; inhibit photosynthesis in plants (already in deep trouble from a lack of visible light); suppress the normal processes by which DNA molecules are repaired in living organisms; and cause corneal damage in the highly developed eyes of mammals. The same sorts of molecules that removed the ozone would also settle as toxic air pollutants, inhibiting the recovery of any plant and animal species that did survive the darkness.

For some would surely survive, though in much-diminished numbers. A few months of darkness would not, for instance, harm the seeds of most plants that grow in temperate climates, so they would have the chance to sprout and to reestablish their species. In addition, many temperate plants have built-in mechanisms to survive through periods of low sunlight and intense cold. In contrast, tropical plants typically have no such mechanisms, and their seeds cannot survive for long periods, as temperate-zone plant seeds can.

Even such general statements arouse some controversy among biologists, and when it comes to predicting just which species are most likely to face extinction, we have little to go on but opposing ideas. Warm-blooded mammals, who need to eat more than cold-blooded reptiles, seem at a disadvantage under conditions of extreme cold; on the other hand, some mammals can hibernate quite well for months on end, reducing their internal temperatures and their need for food. A debate now exists between those who believe dinosaurs to have been "standard" cold-blooded reptiles and those who point to the enormous sizes of many dinosaurs as evidence that the standard reptilian approach could not work, and that dinosaurs must have been (more or less) warm-blooded, not simply responsive to the ambient temperature but capable of significant self-regulation of their internal temperatures. The long and short of this debate is that so far as the extinction of the dinosaurs is concerned, we have no single characteristic at which we can point and say, "That is why the dinosaurs died while our mammalian ancestors survived." Nevertheless, it seems reasonable to assume that a planetwide catastrophe which extinguished many forms of life could have taken the dinosaurs as well.

Within three to six months after the dust from an impact first shrouded our atmosphere, the earth's surface temperature would gradually rise to its normal level. Within a decade or two, many plants would again grow tall, albeit in a highly altered ecosystem of reduced diversity. Phytoplankton that managed to survive the darkness would quickly come to dominate the oceans, and any surviving species of animals that fed upon them would likewise flourish. Best off would be the benthic (bottom-dwelling) forms of sea life, accustomed to living without much light and buffered against temperature changes by the upper layers of the ocean. Benthic communities do depend for food on matter drifting toward them from higher levels and so are not completely independent of the upper layers, but corpses would be in plentiful supply after a period of darkness. Benthic species might therefore take over ecological niches vacated by newly extinct species that used to flourish above them. In the most extreme scenarios we can imagine, the benthic communities represent Earth's secret weapon against the total destruction of living creatures: From the ocean depths could arise the next forms of intelligent life, so that the sharks of today, which have already survived for two hundred million years with few changes, could evolve into the world leaders of tomorrow.

The preceding scenario gives our best estimate of the effect of injecting several billion tons of dust into the atmosphere. Whatever the actual outcome, no one doubts that the event would spell doom to many (though not all) species. The vertebrate animals—mammals in particular—would suffer tremendously, since they are highly sensitive to ultraviolet radiation, depend on sunlight to find their food, and cannot store their germ plasm in the form of seeds: Even the eggs of birds and reptiles need warmth to survive.

REBUTTAL: Alternative Explanation of Mass Extinctions

Cosmic Rays: Too Far Out?

Critics of the impact theory of mass extinctions, when challenged to produce other explanations of catastrophes such as the Cretaceous-Tertiary extinction, have produced two different hypotheses, one poor, the other quite reasonable.

The unlikely hypothesis blames cosmic rays for the destruction of many species of plants and animals on Earth. Cosmic rays, misleadingly named before physicists discovered their true nature, are not "rays" but elementary particles—electrons, positrons, protons, and other nuclei—moving at nearly the speed of light, and reaching Earth in nearly equal amounts from all directions. Earth's magnetic field affects the trajectories of virtually all the electrically charged particles in the cosmic rays. This field creates (imaginary) lines of magnetic force, which spread around the world from the north to the south magnetic poles, giving the earth a "halo" of magnetic force that surrounds our planet as if it were an apple core protected by magnetic outer layers. As charged particles approach the earth, our magnetic field deflects many of them into new trajectories, in which the particles spiral for several days around one of the lines of magnetic force before eventually reaching the atmosphere. The particles trapped by the magnetic field at any given time constitute the Van Allen belts high above the earth.

At intervals of a few million years, the magnetic field reverses its polarity, changing its direction of magnetization by 180 degrees. When this occurs, the strength of the field temporarily falls close to zero. Some scientists have speculated that at such junctures, the flow of cosmic-ray particles to the earth's surface increases dramatically. The fast-moving particles can cause mutations if they strike the chromosomal material of a living cell, and a large flux of such particles can kill entire organisms and so might cause periodic extinctions.

Unfortunately, this theory fails to take account of all the facts. The earth's magnetic field has little effect on cosmic-ray particles approaching along trajectories almost directly above the north and south magnetic poles. Yet measurements show that the flow of cosmic rays at the magnetic poles rises no higher than the flow in Europe or North America, and life near the magnetic poles shows no signs of ill effects that might be caused by cosmic rays. Therefore we cannot conclude that times of low magnetic field allow cosmic rays to have important effects for life on Earth.

A better variant of the cosmic-ray hypothesis, proposed by the late Soviet astrophysicist Josef Shklovskii, among others, suggests that relatively nearby exploding stars could flood the Earth with tremendous amounts of cosmic-ray particles, killing many forms of life. Since such explosions, called supernovas, are the most likely origin of cosmic rays, the theory has merit. We cannot estimate exactly the number of cosmic-ray particles that would be created by a supernova outburst, nor can we determine just how strong a flood of these particles would produce a mass extinction without totally eliminating all life on Earth. Rough estimates suggest that if a star within 50 light years of the sun—one of the thousand or so nearest stars— were to become a supernova, its cosmic-ray flux would be sufficient to produce catastrophic effects. The closest supernova to ourselves ever observed was 5,000 light years away, and apparently had no effect on life on Earth when its cosmic rays arrived about 900 years ago.

On the average, we can expect one supernova explosion among every million stars about once in every ten million years. Among one thousand stars, we would therefore expect to find a supernova explosion only once in every ten *billion* years—close to the age of the Milky Way. This low probability argues against the supernova explosion for *any* of the mass extinctions during the past 250 million years, and certainly would eliminate the possibility that cosmic rays from supernova explosions can explain all of the mass extinctions, if our estimates of the flow of cosmic-ray particles from supernovas, and the effects of these particles on Earth, are even approximately correct.

So although the supernova-explosion hypothesis has a certain appeal, it loses plausibility upon closer examination of the numbers. The theory definitely predicts no periodicity whatsoever in mass extinctions, since we believe that supernova explosions are random, independent events connected only with a star's internal evolution. In any case, in view of the grave uncertainties connected with the cosmic-ray theory, we should be wary of accepting this hypothesis without careful examination of competing ones.

Terrestrial Explanations for Mass Extinctions

The most promising alternative to the hypotheses for mass extinction that look far beyond Earth, to cometary or asteroidal impacts or to a flood of cosmic-ray particles, consists of the more mundane notion that significant changes in terrestrial conditions—quite possibly unrelated to anything outside the earth—have had a significant effect on life. This explanation has the great advantage of resting on the well-established facts that organisms are extremely sensitive to environmental changes, that such changes are well documented in the geological record, and that the mass extinctions, especially those at the Permian-Triassic and Cretaceous-Tertiary boundaries, are known to have occurred during times of significant changes in global environments.

The end of the Cretaceous was indeed a time when the marshes dried up and the inland seas—for example, the sea that ran through the western center of what is now the United States—largely receded. Harvard University's renowned biologist Stephen Jay Gould, who is impressed by the evidence for the 26-million-year period in the mass-extinction data, has written that "for more than one hundred years, geologists have sought terrestrial agents to associate with mass extinction. The litany is long, yet all but one have failed—mountain building, volcanism, fluctuations in temperature, to name just a few old and unsuccessful favorites. Falling sea level represents the one good correlation (and the 26-million-year-cycle theorists had better take it into account). Nearly all mass extinctions are preceded by a marked regression of sea level."* A more single-minded critic of the impact theory might suggest simply looking for more detailed links between changes in environment and the survival of living organisms instead of scanning the heavens.

The paleontologist David Jablonski, of the University of Arizona, once Gould's student, has become a leading expert on interpreting the fossil record in a way that takes into account that a falling sea level would cause a decline in the rate at which sedimentary rocks form, and hence an inevitable decline in the volume available to contain fossils. Jablonski has attempted to correct for this phenomenon, which would cause a reduction in fossils whether or not any mass extinction occurred. He has looked for groups of life forms that reappear in the fossil record once the sea level had ceased to fall, and calls these groups *Lazarus taxa*. To separate the effect of a falling sea level from any true mass extinction, Jablonski has determined the number of Lazarus taxa that disappear before the extinction but reappear afterward. This allowed him to estimate the extent to which an apparent decline in the abundance of life forms is merely the result of the lower volume of sedimentary rock available to hold fossils. Some of the declines in the fossil record turn out to be completely explained by the falling-sea-level effect; others clearly are not.

Jablonski's analysis strengthens the argument that *some* mass extinctions—the one at the Cretaceous-Tertiary boundary is a notable example—must have occurred suddenly. We still cannot begin to prove that "suddenly" means during a period of a few years rather than a few thousand, but since sea-level declines typically occur over hundreds of thousands, or even millions, of years, we can say with some confidence that

* From "The Ediacaran Experiment," by Stephen Jay Gould, in *Natural History*, February 1984, Vol. 93, no. 2, pp. 14–23.

worldwide changes in environment cannot explain the largest mass extinctions. If they could, then the apparent periodicity of these extinctions might be simply a coincidence, or might tell us that such environmental changes themselves recur in cycles for as-yet-unknown reasons.

Gould and Jablonski might be described as already partial to notions of abrupt changes in forms of life, for Gould is best known for the theory of "punctuated equilibrium" (alternating periods of the rapid appearance of new species with periods of relatively little change in species), which he developed in collaboration with Niles Eldredge, of the American Museum of Natural History. As an iconoclast, Gould would naturally appreciate the work of Raup and Sepkoski: "If punctuated equilibrium upset traditional expectations," writes Gould, "mass extinction is even worse."* In the final chapter of this book, we shall examine the implications for the evolution of life here and elsewhere of any theory that predicts periodic mass extinctions. Our immediate task, however, must be to make a detailed assessment of the theories that explain such periodic behavior as the result of periodic changes in the earth's cosmic environment, rather than "mere" variations in terrestrial phenomena unlinked to changes in the cosmos.

*Ibid.

4

THE FINAL MYSTERY: WHAT BRINGS THE COMET SHOWERS?

To THE EXTENT that we accept the hypothesis that mass extinctions have recurred on a 26- to 32-million-year cycle, and that these extinctions arise from the periodic occurrence of collisions with one or more comets or asteroids, we have now cleared the decks for the final theoretical confrontation: What induces showers of comets by diverting them toward the planetary region of the solar system?

In view of theoreticians' ability to produce hypotheses that will explain almost any concatenation of events, it may be surprising to learn that only three basic theories of how comets come to shower the inner solar system have been seriously advanced. Let us now examine in turn the two chief theories and the one minor theory, looking for their advantages and drawbacks, in the hope of arriving at some insights that will allow us to look outward at the rest of our galaxy with new ideas about our own planet's history and the evolution of life on it.

The chief competing theories of mass extinction all seek a mechanism to divert comets from the Oort cloud into the inner solar system. All recognize that any such diversion must depend on the gravitational force that a perturbing object exerts on the comets, and all need to find a periodic recurrence of such perturbations in order to explain the 26-to-28- or 30-to-32-million-year cycle that they think they have found in the extinction rate and impact-crater ages. All theories have their backers, who often regard the competing theories as inadequate, contrived, and unsupported by the data. Let us attempt to look with detachment upon this competition of hypotheses, seeking to find the key elements that may allow us to discriminate between more and less accurate representations of reality.

The Minor Theory: Planet X

The theory taken seriously by the fewest astrophysicists, proposed in 1984 by the astronomers Daniel Whitmire and John Matese, of the University of Southwestern Louisiana, hypothesizes that a tenth planet of the sun's family perturbs comets into the inner solar system. Like the Nemesis theorists, who propose an unseen star as the perturbing force, Whitmire and Matese hypothesize a previously undetected object, the often-suggested, never-detected Planet X, whose orbit would lie outside Pluto's orbit, say between fifty and one hundred times the Earth–sun distance (50 to 100 A.U.).

If Planet X exists, and if its orbit has a significant elongation, then at points along its orbit at greater distances from the sun, the planet could perturb those comets whose orbits bring them to within one hundred A.U. of the sun. Such cometary orbits are much smaller than those of typical comets in the Oort cloud, which move around the sun at distances of 10,000 to 40,000 A.U. In order to produce a significant effect on even the comets with smaller orbits, Planet X must have a mass larger than Earth's, and hence much greater than Pluto's (one six-hundredth of Earth's mass). A smaller mass would give Planet X too little gravitational force to produce the orbital perturbations.

The Planet X theory has three parts, the first of which is the speculation that Planet X indeed exists. In part two of their hypothesis, Whitmire and Matese speculate that a large number of comets are to be found at about Planet X's distance from the sun, just outside the planetary region of the solar system. These comets would have been expelled, soon after they formed, by Neptune's and Uranus's gravitational force, but to much lesser distances from the sun than most of the comets, which collectively form the Oort cloud.

The third part of the Planet X hypothesis holds that the planet would have an orbital period of only about 1,000 years, so that we must look further to produce comet showers every 26 to 32 million years. Whitmire and Matese propose that Planet X moves on a highly elongated orbit that slowly precesses—changes its orientation in space—over a period of this length. Planet X's gravitational force tends to create a "clear zone" at the inner region of the Oort cloud, expelling comets from that area, but as the planet's orbit precesses, it eventually brings the planet close to the edge of the clear zone, where numerous comets are still to be found. At these times, 26 to 32 million years apart, the tenth planet can divert large numbers of comets into orbits that penetrate the inner regions of the solar system.

The triple assumption of the Whitmire-Matese model—a hitherto undetected planet; a new belt of cometary orbits, smaller than the Oort cloud; and an elongated, precessing orbit for Planet X—has proved too far-fetched for most astronomers to accept. This does not prove the Planet X theory wrong; it merely reminds us of the several different assumptions contained in the hypothesis.

The Major Theories: Nemesis and Galactic Oscillation

The real competition occurs between the Nemesis theory—that a solar companion star produces comet showers at 26-million-year intervals—and the theory that the sun's up-and-down motion, which carries the sun through the median plane of the Milky Way every 33 million years, produces comet showers as the sun passes close to interstellar clouds of gas and

Figure 8—The spiral galaxy M81, which we observe nearly face-on, closely resembles our Milky Way. M81's spiral arms mark the regions where the youngest, most luminous stars have formed within the past 10 million years. (Lick Observatory photograph.)

dust. These clouds concentrate near the median plane (the imaginary plane that divides the disk-shaped galaxy into top and bottom halves), so the probability of a close encounter with such an interstellar cloud increases at intervals of 33 million years. Each cloud exerts enough gravitational force on the solar system to perturb comets into the inner solar system, thus producing a comet shower. This concept can be described as the galactic-oscillation theory of mass extinctions.

Our home galaxy, the Milky Way, contains the sun and several hundred billion other stars. We can picture the Milky Way as an enormous, highly flattened disk, with a bulge at its center, packed with extraordinary large numbers of stars, that astronomers call the galactic nucleus. The disk has a diameter of about 100,000 light years, and the solar system, 30,000 light years from the center, lies in the galactic "suburbs." To send a radio message to the galactic center and to receive an answer would take 60,000 years, a reminder that the galactic disk has a diameter just about a billion times larger than Pluto's orbit around the sun, and more than two thousand times the distance from the sun to its closest neighbor stars. If we could fit the planets' orbits around the sun onto a platter, the sun's nearest neighbors would be a few kilometers distant, and the Milky Way would span the United States.

The vast disk of the Milky Way consists mainly of stars, plus an important sprinkling of interstellar gas and dust among the stars. Averaged over the entire galaxy, this interstellar matter amounts to about one atom in every cubic centimeter—a density that ordinary water exceeds by a factor of one trillion trillion. Some interstellar matter is found throughout the galactic disk, but most of it appears in clumps called interstellar clouds, regions of interstellar space with densities of matter much greater than average.

Many such interstellar clouds have sufficiently large densities for the atoms within them to collide and form molecules, so the matter there exists primarily in molecular form. These molecular clouds typically contain one to ten thousand times the sun's mass, mostly in the form of hydrogen (H_2) molecules, with a smaller amount of mass in other molecules and in much larger (but still-tiny) dust particles. Some of the molecular clouds have 1 to 10 *million* times the mass of the sun. These giant molecular clouds represent places where stars will soon be born, or are being born, as clumps within the cloud draw together under the influence of their own gravitational forces.

The closest giant molecular cloud lies in the direction of the constellation Orion, the Hunter, at a distance of about 1,600 light years. This cloud, about 100 light years in diameter, contains a region within which stars have formed recently—in astronomical terms—during the past million years. These new-born stars represent a star cluster in the making, and shine as the middle "star" in the sword of Orion. The giant molecular cloud that contains them can be studied by the radio waves that its various molecular constituents emit or absorb. From such study, astronomers have deduced much about the com-position, mass, and age of the cloud, and have extrapolated their findings to make conclusions about the other molecular clouds strewn through the disk of the Milky Way.

Crossing the Median Plane

The salient fact about molecular clouds, for our purposes, is their concentration toward the median plane of our galaxy. Because of this concentration of molecular clouds, the chance of diverting comets to produce a comet shower increases as we approach the plane. The Milky Way's disk spans 100,000 light years, but its thickness amounts to no more than one or two thousand light years. The system of molecular clouds, which in aggregate contain only a few percent of the total mass in the disk, tend to lie within only a few hundred light years of the median plane. At a distance of 250 light years above or below

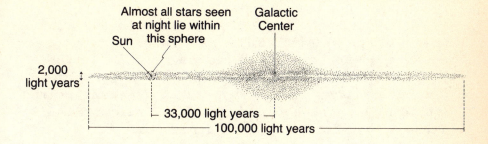

Almost all stars seen
at night lie within
Sun this sphere

Galactic
Center

2,000
light years

33,000 light years

100,000 light years

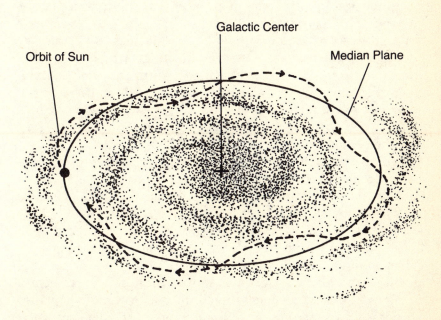

Orbit of Sun

Galactic Center

Median Plane

Figure 9—A drawing of a side view of the Milky Way shows that the sun lies almost exactly at the galaxy's median plane, the plane of symmetry within the galatic disk. As the sun orbits the galactic center, it oscillates up and down through the median plane, reaching distances of about 200 light years on either side. (Top: Marjorie Baird Garlin. Bottom: David Wool.)

the median plane, there are only half as many molecular clouds per unit volume as at the median plane.

All of the mass in our giant spiral galaxy, distributed in a shape like a pie plate but a trillion trillion times larger, attracts the sun and the entire solar system through gravitational force. When we calculate the combined effect of the gravitational forces from all parts of the galaxy, we find that their net force is equivalent to the force exerted by a single imaginary object, located at the galactic center, with a mass of about 200 billion solar masses.

Pulled by this force, the sun and its planets move around the center of the Milky Way in an orbit two billion times larger than the Earth's orbit around the sun. Like the planets in their orbits, the solar system in its orbit balances the effect of gravitation, which attracts us toward the galactic center, with momentum in orbit, loosely called centrifugal force. As a result, the sun and its planets neither fall into the galactic center nor move farther away from it; instead, they move in a nearly circular orbit *around* the galactic center, 30,000 light years from the center, taking 240 million years per orbit.

The solar system not only orbits the galactic center in a nearly circular orbit but, as it does so, also bobs up and down through the galactic plane in an oscillatory motion. This motion carries the sun 200 light years above and below the plane as it makes its 100,000-light-year circuit of the galactic center. As the sun rises from the plane, the total mass of the galaxy acts to pull it back toward it, so the sun's motion in this direction falls to zero. At its maximum distance above or below the plane, the sun reverses direction, falls toward the plane, and passes through it because of its momentum. The sun then emerges on the other side of the median plane, where the same effect occurs. The sun and its planets apparently formed from a clump of matter that was moving perpendicular to the median plane as well as around the galactic center. The solar system preserves this motion, and bobs through the plane like a horse on a carousel, making about seven passages through the plane in each orbit. The period between successive passages equals 33 million years (possibly a bit longer), and we are now so close to the median plane that we can say that we are either "in" such a passage, or just emerging from one. Although the period between crossings of the plane might possibly be only 31 million years, astronomers state with confidence that the period simply cannot be 28 million, let alone 26 million, years. Hence the galactic-oscillation theory does not permit the interpretation of the mass-extinction and impact-crater data made by Alvarez and Muller (along with Marc Davis and Piet Hut, of the University of California at Berkeley); if you believe in the shorter period (26 million, rather than 33 million, years) for these events, you are driven toward the Nemesis hypothesis.

Molecular Clouds and Their Perturbations of Comets

The concentration of molecular clouds toward the galactic median plane makes it tempting to identify the 33-million-year interval between successive passages through the plane with the period between successive impacts on Earth. The molecular clouds provide a mechanism to produce such impacts: Their large masses, up to a million times the sun's mass, imply that they exert large gravitational forces, even though they consist of nothing but gas and dust. Such gravitational forces can divert comets from their normal orbits around the sun, even if the molecular cloud approaches the sun to a distance of "only" a few tens of light years, rather than the closer distances of approach that would be required for a less massive object to produce similar perturbations. In 1984, Michael Rampino and Richard Stothers developed this theory in some detail, analyzing the mass-extinction and impact-crater data and finding cyclical patterns with periods ranging from 30 to 33 million years.

The British astronomers Victor Clube, of Oxford University, and William Napier, of the Royal Observatory at Edinburgh, have also explored this theory, calculating the perturbations that a molecular cloud would produce on cometary orbits. They agree with Rampino and Stothers that molecular clouds are the most likely source of the perturbations of comet orbits; that these perturbations tend to occur when the sun crosses through the Milky Way's median plane; and that the average interval between successive perturbations must therefore be the 33 million years between successive passages through the plane.

On the other hand, two American astrophysicists, Patrick Thaddeus and Gary Chanan, of Columbia University, have made an analysis that strongly refutes this hypothesis. Thaddeus and Chanan agree that the number of molecular clouds per million cubic light years is greatest at the galaxy's median plane, but they point out that molecular clouds also appear at distances within a few hundred light years above and below the plane. The giant molecular cloud in Orion, for example, lies nearly 400 light years from the plane of the Milky Way. Thaddeus and Chanan point out that since the sun's up-and-down oscillation carries it only 200 light years above and below the galactic plane, even though close encounters with molecular clouds will grow more frequent as the sun passes through the plane, *some* close encounters can occur at any time. In other words, although molecular clouds do concentrate toward the median plane, their concentration is not strong enough to produce a highly variable effect on the sun as it moves only a relatively small distance above and below the plane. Cometary showers produced by close encounters with molecular clouds would not show sharp peaks at 33-million-year intervals in the biological or geological record on Earth. Instead, we might expect cometary showers at any time, with only a mild increase as the sun passes through the galaxy's median plane.

To use an exaggerated analogy, Thaddeus and Chanan are

suggesting that the galactic-oscillation explanation could be compared to a carousel rider's search for variations in the air pressure caused by altitude changes along his path: They exist but have no significant effect. We may note that because close encounters can occur at any point in the sun's up-and-down motion, the galactic-oscillation theory allows for some "slip-page" in the match between mass extinctions and crossings of the plane, and in fact, this could help to explain the imperfect match between the idealized cycle and the actual data on mass extinctions and impact craters. Thaddeus and Chanan would reply that the molecular clouds' lack of strong concentration toward the plane is so noticeable that the theory predicts all "slippage" and no "peaks"; we should expect to find no "cycles" whatsoever, only random occurrences of large impacts and mass extinctions!

To its credit, the explanation of comet showers proposed by Rampino and Stothers, and by Napier and Clube, has one great appeal for scientists: It relies on objects (molecular clouds) and phenomena (gravitational perturbations of cometary orbits) that definitely exist. Of course, this certainty could eliminate the theory, since we also know the period between passages through the median plane: 33 million years, with a possible error of one or two million years. If closer analysis of the extinction and impact data points more strongly toward this period, the galactic-oscillation explanation will move toward wider acceptance; if such analysis tends to suggest a period close to 26 million years, the Nemesis theory will ascend; and if further analysis suggests that the data exhibit no real periodicity, we shall hear considerably less about both hypotheses.

We do know that relatively close encounters of the solar system with molecular clouds must occur, but we do not know just how many comets they may divert, nor how close an approach is required to produce a particular effect. We can be sure that the molecular clouds *are* responsible for diverting some comets into smaller orbits, but we do not know, for example, whether they play a more important or less important role than individual stars that happen to pass within a few light years of the sun.

Unknown Matter in the Galactic Plane

There is a way to modify the galactic-oscillation theory that answers Thaddeus and Chanan's objection, though at the price of hypothesizing a new component, of unknown form, of the Milky Way. Astronomers can calculate, by studying the up-and-down oscillations of stars, how much matter lies close to our galaxy's median plane. These calculations show that *stars*, the matter we can easily identify, form only about half of the total matter. Interstellar clouds plus diffuse interstellar matter have less total mass than the mass in stars, so some of the matter in our galaxy—even the parts between ourselves and the closest stars—seems to be "missing," unaccounted for in any known type of matter.

Intriguing speculations about what form such matter might take—burnt-out stars, rocks, black holes, or even more exotic hypothesized forms called particle strings (completely unlike any familiar forms of matter) testify to the hard work and vivid imaginations of theoretical astrophysicists. What counts for now is the fact that the "missing matter" apparently does exist, even though we have yet to observe it directly, and know of its existence only from the "fingerprint" of the gravitational force that it adds to the total. This matter might be concentrated toward the median plane even more sharply than the distribution of stars or of molecular clouds. Indeed, we might find that our galaxy has "rings" of material, analogous to the famous rings of Saturn, but ten trillion times larger. Matter in these rings would orbit the galactic center, but would never rise more than a fraction of a light year above and below the median plane.

If this is true, then every 33 million years the total gravitational force from this matter on the solar system would suddenly change direction, from "upward" to "downward" (or vice versa) as the sun passed through the Milky Way's median plane. This change of direction of the gravitational force from unseen matter could perturb comets into new orbits and could produce a comet shower, sending billions of comets toward the planetary region of the solar system. In this case, the galactic-oscillation model would seem valid because the unseen matter, by hypothesis, concentrates with almost complete sharpness at the median plane, instead of diffusing somewhat above and below it like the molecular clouds.

Since we have yet to detect the form and location of the unseen matter in our galaxy, we cannot tell how strongly this matter concentrates toward the plane, and therefore must simply theorize that this concentration is extremely strong. This places the unseen-matter variation of the galactic-oscillation theory on a footing similar to that for the Nemesis theory: We have invented an object (or, in this case, part of the Milky Way galaxy) that has the properties we need to make our theory satisfy the period seen in the data. Until we find the "missing matter" and determine its distribution in the Milky Way, our attempt to "save the hypothesis" of the galactic-oscillation theory must remain a speculation—one worthy of the great game that we play in our search for a viable model of what we observe.

The Nemesis Theory: The "Death Star"

The competing Nemesis theory uses the same, already known phenomenon (gravitation) to explain cometary perturbations that the galactic-carousel theory does, but it relies on an entirely unknown object to cause the perturbations. Hence, the theory can assign the Nemesis star the most effective orbit to produce the observed results, which the Nemesis theorists insist is a period close to 26 million years in the extinction rate

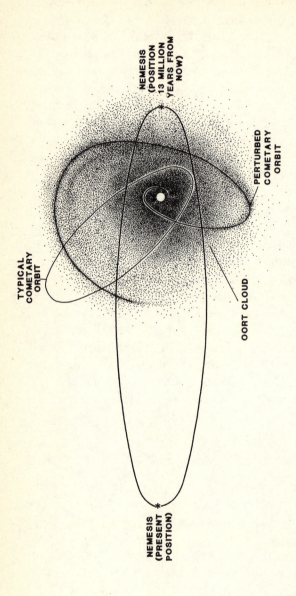

TYPICAL
COMETARY
ORBIT

NEMESIS
(POSITION
13 MILLION
YEARS FROM
NOW)

PERTURBED
COMETARY
ORBIT

OORT CLOUD

NEMESIS
(PRESENT
POSITION)

Figure 10—The Nemesis theory of cometary perturbation hypothesizes that there exists a solar companion star which moves in an elongated orbit, carrying it from distances of 25,000 A.U. (Earth-sun distances) out to 150,000 A.U. from the sun. At the closer points of its orbit, Nemesis could perturb the orbits of billions of comets in the Oort cloud, which orbit the sun by the trillions at distances of tens of thousands of A.U. Some of these perturbed comets would enter the planetary region (0.4 to 40 A.U. from the sun) of the solar system, and a tiny fraction of them might strike the earth. The white disk at the center of the illustration represents regions of the solar system (including the sun and its planets) that are inside the Oort cloud. (Drawing by Augusta Lucas-Andreae.)

and close to 28.4 million years in the impact-crater record. Skeptics may point out that this freedom of hypothesis speaks against, not for, any theory, making it more contrived (*ad hoc* is the more genteel scientific term) than one can accept. In addition, the skeptics say, the Nemesis theory requires that the sun be a member of an unusual star system, a double-star system in which a low-mass companion moves in an extremely large orbit around the system's center of mass, which would be close to the much more massive star in the system, our sun. Such a double-star system would be unlike almost all the stars and star systems that we observe around us.

Since the growth of astronomical knowledge has always led us toward an increasing belief in our "averageness," we should be leery of adopting any theory that contradicts our hard-won experience. Before we examine the ways in which the Nemesis debate could be resolved, we shall try to determine whether the Nemesis theory, honed as finely as its supporters can, proves capable of satisfying both the data and the requirements imposed by Newton's laws of gravitation and of motion. Only by demonstrating that Nemesis's purported behavior violates no fundamental laws of physics can we begin to establish a credible hypothesis.

The Orbit of Nemesis Around the Sun

If we seek to explore the notion that a solar companion star, Nemesis, produces comet showers at 26-million-year intervals, the most immediate and logical cornerstone of the hypothesis would be that Nemesis reaches the closest point in an elongated orbit around the sun every 26 million years. If Nemesis does exist, then Nemesis and the sun must orbit around their common "center of mass." If Nemesis had the sun's mass, their common center of mass would lie halfway between the two stars, and both stars would move in orbits of equal size. However, we can calculate that the sun must have at least eight times the mass of Nemesis, or else Nemesis would be so luminous that we would surely have already detected it. Hence the center of mass of the sun-Nemesis system must lie between the two objects, but at least eight times farther from Nemesis than from the sun. This means that we can approximate the proposed sun-Nemesis system by imagining that the sun remains stationary while Nemesis orbits around it. Of course, both Nemesis and the sun, along with the sun's planets, asteroids, and comets, would orbit the center of the Milky Way, bobbing up and down through the galaxy's median plane.

It is a remarkable fact, discovered by Johannes Kepler and demonstrated mathematically by Sir Isaac Newton, that when a less massive object moves around a more massive object in response to the latter's gravitational force, the less massive object will move along an elliptical (or circular) trajectory. Furthermore, the orbital period of any such object will depend only on the mass of the more massive object and on the length of the long axis of the elliptical orbit. The *shape* of the orbit—extremely elongated or nearly circular—does not matter at all in determining the orbital period; only the length of the long axis counts.

Kepler and Newton showed that the orbital period varies as the $3/2$ power (the square root of the cube) of the ellipse's long, or major, axis. (The minor axis is the short distance across the ellipse, bisecting the long axis.) Earth's slightly elliptical orbit has a major axis equal to the long diameter of the orbit, or just over 2 A.U. Since the $3/2$ power of 4 equals 8 ($4^3 = 64$; $\sqrt{64} = 8$), we can immediately calculate that an object whose orbital major axis equals 4 times that of Earth (as is true for some asteroids) will take 8 years for each orbit. An orbit with a major axis 100 times Earth's would have an orbital period of 1,000 years, whether the orbital shape resembled a circle or the outline of a football.

Every elliptical orbit has two foci located along its major axis, equidistant from the axis's midpoint. An ellipse is a curve determined by the restriction that at every point along the ellipse, the sum of the distances to the two foci remains constant. For a circle, the two foci coincide, and the "sum of the distances" is simply twice the radius (the diameter) of the circle. In the solar system, the sun occupies one focus of all the planets' elliptical orbits, but the other focus is simply a point in space.

Newton's mathematics also shows that any object with an orbital period of 26 million years must move in an elliptical orbit whose major axis is 88,000 times the diameter of the Earth's orbit, or 176,000 A.U., equal to almost 3 light years. Nemesis, the proposed solar companion star, must therefore have this major axis for its orbit. If its orbit were circular, Nemesis would always be the same distance from the sun: half of the major axis, or 1.5 light years. In this case we would expect Nemesis to produce no comet showers at all. If we want Nemesis to perturb comets (and this is the whole reason for hypothesizing such a "death star"), we must therefore assume that Nemesis has a noticeably elliptical (elongated) orbit. Then Nemesis's closer approaches to the sun—and to the Oort cloud—would produce much greater-than-average perturbations of the comets' orbits, and would induce comet showers in the inner solar system.

Calculations have been made to determine just *how* elongated Nemesis's orbit would have to be to produce the desired effect. Nemesis would move in a highly, but not tremendously, elongated orbit, one in which its distance from the sun falls to a minimum of 26,000 A.U. and increases to a maximum of 150,000 A.U. (Notice that the sum of these distances is just the major axis of the orbit.) Since the force of gravity increases in proportion to one over the *square* of the distance between two objects, Nemesis at its close approaches to the sun would have a far stronger effect on comets in the Oort cloud, most of which are from 10,000 to 40,000 A.U. from the sun, than it would at its average distance (88,000 A.U.), let alone at its maximum distances from the sun.

The Nemesis theorists Marc Davis, Piet Hut, and Richard

Muller were therefore on firm ground when they claimed that
a solar companion with such an orbit could easily produce a
peak at 26-million-year intervals in the rate of cometary diver-
sions. Since Nemesis moves more rapidly as it approaches the
sun, we can calculate that the part of its orbit that brings
Nemesis closer to the sun than 40,000 A.U. lasts for only about
one million years. During this million years, we would expect a
tremendous increase in the number of comets diverted toward
the planetary regions of the solar system.

The Nemesis theory, as derived from the conclusion that a
26-million-year cycle does exist, requires Nemesis to have an
orbit with the size and shape described above. Too elongated
an orbit would cause Nemesis itself to have an unstable, fast-
changing orbit, so that the periodicity would be lost. On the
other hand, an orbit that was nearly circular could not provide
a sufficient peak, or "pulse," during the million years or so
when Nemesis would be closest to the sun, because its gravitational force on the comets in the Oort cloud would grow only
slightly stronger than average, since Nemesis would be only
slightly closer than average to the sun and its comets.

The Nemesis theorists can thus predict not only the major
axis of their star's orbit (176,000 A.U.) but also the amount of
elongation (not too circular, not so highly elongated that the
orbit would be unstable) of that orbit. Furthermore, the proponents of Nemesis can make a good estimate of the *mass* of the
hypothesized solar companion. This estimate sets upper and
lower limits on Nemesis's mass—an upper limit based on the
fact that we have not yet seen Nemesis, and a lower limit based
on the reason for the Nemesis hypothesis: the perturbation of
large numbers of cometary orbits.

The Effect of Mass on a Star's Luminosity

The upper mass limit on Nemesis derives from our knowledge of how stars shine. All stars produce energy deep in their

interiors by fusing hydrogen nuclei (protons) into helium nuclei. Each such nuclear-fusion reaction turns energy of mass—the energy packed into any particle with mass, simply by virtue of its existence—into kinetic energy (energy of motion), the sort of energy "useful" to us. Kinetic energy from the sun streams into space as light, infrared and ultraviolet radiation. This energy is a product of hydrogen-to-helium fusion, which is duplicated in the workings of a hydrogen bomb, but on a scale that makes the latter look like a drop in the ocean: Our sun produces more kinetic energy every second than a billion hydrogen bombs!

The more massive a star is, the more kinetic energy it produces. The rate of energy production varies roughly as the *cube* of the star's mass. Rigel, a supergiant star with 20 times the sun's mass, produces 10,000 times more energy per second than the sun; only Rigel's enormous distance from Earth, 40 million times the sun's, makes Rigel appear as just another star in the winter skies. In contrast, Barnard's Star, which has about one-tenth of the sun's mass, produces only about one two-thousandth of the sun's energy each second. Because of this low luminosity (energy output per second), Barnard's Star, even though it is the fourth closest star to the sun, does not rank among the top hundred thousand stars in its apparent brightness. Rigel, more than a hundred times farther from us than Barnard's Star, seems far brighter, a tribute to the tremendous amount by which Rigel's rate of energy production exceeds that of Barnard's Star.

How Massive Is Nemesis?

For Nemesis to have escaped recognition as the closest star to the sun, its luminosity would have to be no greater than about three times that of Barnard's Star. A star's apparent brightness depends both on its luminosity and on its distance from us. The Nemesis theory puts the last peak in the extinction pattern about 13 million years—half a cycle—into the past, which implies that Nemesis now occupies one of the far points of its orbit. Since we know the major axis and the elongation of Nemesis's proposed orbit, we can calculate that at its maximum distance, Nemesis must be about 150,000 A.U. from the sun, or 2.6 light years away. Barnard's Star, at a distance of 6 light years, would be just over twice as far away as Nemesis at its current distance.

Since an object's apparent brightness decreases in proportion to the square of the distance from the source to the observer, we can proceed to calculate that if Nemesis and Barnard's Star have the same luminosity (rate of energy production), then Nemesis would have more than four times the apparent brightness of Barnard's Star. If Nemesis has a luminosity 3 times that of Barnard's Star, its apparent brightness would be 12 times greater, and we would almost certainly have detected it by now in one of the sky surveys that astronomers have made, trying to find stars relatively close to the sun by studying the likely candidates one by one. From our knowledge of how a star's luminosity depends on its mass, we may therefore assign Nemesis a maximum mass no greater than 1.2 times the mass of Barnard's star, or about 0.12 solar masses.

We base the minimum mass for Nemesis on the assumption that Nemesis could perturb enough comets to produce the kind of cometary showers hypothesized—ones in which a large comet is likely to strike Earth. This lower limit on Nemesis's mass equals about 5 times the mass of Jupiter, the largest planet. A lesser mass for Nemesis would sharply reduce its ability to divert comets. Jupiter has a mass 318 times Earth's and one-thousandth of the sun's, so we can calculate that Nemesis's minimum mass equals about one two-hundredth of the sun's mass.

Our analysis therefore confines Nemesis's possible mass to a range over a factor of 24, from 0.12 solar masses down to 0.005 solar masses. At the low end of this mass range, for masses less than about 0.05 times the sun's, Nemesis would not be a true star; it would have too little mass for nuclear fusion to occur in its interior (the gravitational force on each part from all the other parts would not "squeeze" the center to the point where nuclei could fuse together). Nemesis would then be what is called a brown dwarf, slowly contracting and emitting infrared radiation as it does so. If this is so—and it is quite reasonable that such low-mass brown dwarfs should exist—then Nemesis would prove much more difficult to discover than if it were a true star, even one with the low luminosity expected for stars of 0.05 to 0.12 solar masses.

Perturbations of Nemesis?

Those who debate the existence of the solar-companion star have been quick to realize that if Nemesis exists, then not only would it perturb comets but also, like a comet, Nemesis would *be* perturbed by the relatively close passage of other stars in the Milky Way, and by close encounters with molecular clouds.

Most of the Milky Way galaxy's hundreds of billions of stars, move, like the sun, in almost-circular orbits around the galactic center, and in the same direction—collectively forming an extremely flat disk. But the stars' orbits are not perfect circles, so as the sun and its closer neighbors perform their 240-million-year motions around the center of the Milky Way, other individual stars may easily approach the sun to within a few light years. The closest star system to the sun, the Alpha Centauri system, consists of three stars, all of which are about 270,000 A.U. from the sun. At this distance, the Alpha Centauri stars have only modest effects on the Oort cloud of comets, which orbit the sun at about one-tenth the distance to Alpha Centauri. But those distant stars *would* have an effect on Nemesis, especially at the far reaches of its orbit. If these far points happened to lie in the direction of the Alpha Centauri system, then when Nemesis was 150,000 A.U. from the sun, it would be as close to Alpha Centauri as it was to the sun. Since the Alpha Centauri system contains more mass than the sun, this would leave Nemesis a likely candidate to become completely "unbound" from the sun's gravitational yoke. Nemesis would then either join the Alpha Centauri system or would go its own way through the galaxy, perhaps later to join another star system.

But it is highly unlikely that Nemesis's orbit takes it to the greatest distances from the sun in just the direction of Alpha Centauri. If, instead, these farthest points from the sun lay somewhat, but not exactly, toward the Alpha Centauri system, those stars would tend to enlarge Nemesis's orbit, without changing it so completely that Nemesis escaped from the sun. Similarly, other stars passing within a few hundred thousand A.U. of the sun would also enlarge the orbit of Nemesis. So, too, would molecular clouds, each with a few thousand solar masses, that passed within a few million A.U. of the sun.

Almost all encounters with stars and molecular clouds would tend to give Nemesis a larger orbit. Eventually they would free Nemesis from the sun completely. The Nemesis theory therefore proposes that when Nemesis and the sun formed, Nemesis moved in an orbit much closer to the sun than it does now. Perturbations have steadily enlarged its orbit, and we happen to exist during the last half billion years or so of Nemesis's gravitational attachment to the solar system. Thus (and skeptics say this decreases the likelihood that the theory is correct), the Nemesis-sun system would represent the unusual situation of a double-star system about to be disrupted (using astronomical time scales). The notion that Nemesis did not form with the sun but was later captured as it passed close by our star seems too unlikely for serious consideration. We therefore theorize that Nemesis is in the last tenth of its lifetime as a member of the solar system—unlikely, but hardly impossible.

The perturbations that would move any solar companion

into a larger orbit would also change its orbital *period*, the time to complete a single orbit. The general trend would be to increase the period as the orbit expanded, but each random change might increase or decrease the period. *Some* change in the period might be just what the theory ordered, in view of the imperfection in periodicity derived from the mass-extinction data. If Nemesis's orbital period changed, more or less at random, by 1 to 3 million years in every orbit, then the data would fit better (admittedly in an ad-hoc way) to the hypothesized orbit of the hypothesized star. Eventually, the period would become "unstable," as Nemesis drifted so far from the sun that the next close encounter would spring it loose entirely.

Critics of the theory have asserted that the Nemesis supporters demand too much when they assume that the sun has an undetected companion *and* that we observe it toward the end of its period of gravitational bondage to the sun. This objection has merit, but Nemesis might yet turn out to be exist and indeed to be just near the end of its gravitational attachment to the solar system. Proxima Centauri, the dim star in the triple Alpha Centauri system, is about 15,000 A.U. from the more massive stars Alpha Centauri A and Alpha Centauri B. This distance gives Proxima Centauri one of the "widest" separations known for any double- or multiple-star system, yet it equals only about one-tenth of the hypothesized maximum distance of Nemesis from the sun, and implies that Proxima Centauri is much more tightly bound by gravitation to *its* major stars then Nemesis would be bound to the sun.

Nemesis theorists would reply that it is extremely difficult for astronomers to detect "wide" double-star systems, since the relatively great separation of the stars means that they move only slowly around one another. Therefore many Nemesis-like systems could exist even among rather nearby stars, yet astronomers would not have found any of them—the more so if the companion "stars" are brown dwarfs, such as Nemesis might be. It will take more than arguments about the statistics of double-star separations to convince scientists to accept, or fully reject, the Nemesis hypothesis.

Where Is Nemesis?

The initial scientific paper proposing the existence of a solar companion star (written by Davis, Hut, and Muller) contained the statement that "if and when the companion is found, we suggest it be named Nemesis, after the Greek goddess who relentlessly persecutes the excessively rich, proud, and powerful. We worry that if the companion is not found, this paper will be our nemesis." Skeptics have found little fault with this concept, and have insisted that until and unless definite proof of Nemesis's existence turns up, the Nemesis theory must be held to be speculation of the wildest sort.

The Nemesis hypothesis has one great test, evident to all who consider it: Find Nemesis! Any theory that explains significant events in the history of Earth through the influence of the nearest star to the sun must rise to the challenge of producing that star, perhaps not immediately, but soon. The Nemesis theory has quite naturally led to the search for Nemesis, which should soon either reveal this star and make more outstanding the careers of its proponents, or fail to produce Nemesis and leave them free to embark on other research, possessors of a glorious idea that did not correspond to reality.

We have seen that the hypothesized solar companion would be a star of extremely low luminosity, the result of Nemesis's small mass in comparison with the sun. That we have not yet detected Nemesis sets an upper limit on its luminosity equal to one eight-hundredth of the sun's. If Nemesis has this luminosity, and a proposed distance from the sun and Earth equal to half the distance from the sun to Alpha Centauri A (whose luminosity equals the sun's), then Nemesis would have an apparent brightness one two-hundredth that of Alpha Centauri A. Nemesis would therefore not rank among the twenty stars of greatest apparent brightness, as Alpha Centauri does, but would be one of the five thousand brightest stars in our sky, too faint to be seen with the naked eye. If, as is quite possible, Nemesis's luminosity equals only one-millionth of the sun's, then Nemesis would be much too faint to be seen without a telescope, only one among many hundreds of thousands of stars recorded on photographic plates. In this case we would have a difficult time finding Nemesis without years of effort. Nemesis might have an even smaller luminosity, which would make our chances of finding the closest star to the sun quite dim.

The Parallax Effect

Whatever Nemesis's luminosity, the best chance of finding the proposed solar companion rests with the parallax effect, the most direct way to measure distances to nearby stars. Earth's motion around the sun gives each star some parallax, an apparent shift in the star's position against the backdrop of much more distant stars. The parallax effect can be observed when you focus on your upright finger at arm's length and close first one eye and then the other: The finger shifts its apparent position against the backdrop of more distant objects. As infants, we investigated this effect without conscious thought, and learned to use the parallax effect with both eyes open, presenting dual images to our brain, in order to estimate the distances to objects, a procedure that our eyes and brains soon performed automatically. We learned then, and can verify now, that *closer* objects show *greater* shifts as the result of the parallax effect. The amount of apparent displacement, the parallax shift, varies in inverse proportion to the object's distance: An object three times farther away than another shows a parallax shift one-third the amount of the closer object's shift.

As Earth orbits the sun, we change our position with respect to our parent star, swinging first 150 million kilometers to one side and then the same distance to the other. We can regard these two positions as analogous to our two eyes, and can observe the parallax shift of nearby objects—in this case stars—against the more distant background, more stars, which exhibit only insignificant parallax shifts because of the stars' tremendous distances from us. All stars are too distant for us to be able to detect their parallax shifts with the naked eye. Such detection requires powerful telescopes, used to photograph stars through the course of a year, so that Earth's motion in space will indeed produce the changes in the stars' apparent positions that we seek to measure. Even with the help of these telescopes, we can measure parallax shifts accurately only for the hundred thousand or so nearest stars, those within about 100 light years of the sun. More distant stars have parallax shifts too small to be determined with accuracy, because the atmospheric blurring sets a limit on the angular displacement that we can measure from the earth's surface.

If Nemesis exists, it should (by hypothesis) be the closest star to the sun. As a result, Nemesis should have the largest parallax of any star, twice as large as that of Alpha Centauri A if Nemesis is half that star's distance from us. No difficulty exists in measuring the parallax; the trick is to find the star that has it. Astronomers have a catalog of the five thousand most likely candidates: the red-dwarf stars, with barely enough mass to escape the brown-dwarf category and to exist as true stars. Nemesis should be such a red dwarf, because its hypothesized low mass (for a star) gives it a low surface temperature, in comparison with the sun and similar stars; most of the visible light emerging from such stars appears at the long-wavelength, red end of the spectrum. Photographic plates have revealed red-dwarf stars by the dozens on every plate, but the hard work of measuring accurate positions at different times during the year has barely begun.

The Nemesis group at Berkeley has started an automated search through the catalog of red-dwarf stars, seeking to determine which, if any, show large parallax shifts. In this search, which began operation in early 1985 and is still being refined in efficiency, a computer-controlled reflecting telescope with a mirror 30 inches in diameter points to one star after another in the catalog, night after night, whenever the sensors at the Leuschner Observatory signal the computer that the skies are clear. In bygone days, photographic plates taken several months apart would be set side by side for examination in a blink comparator, to see whether any stellar images had "jumped" in position between one plate and another. Today a computer records the image in digital form, and no human eye need pass its time looking for jumping images. At full operation, the Leuschner Observatory search should record a thousand stars' positions on every full night of observing. With this ability, a single year should easily reveal whether any of the five thousand brightest red dwarfs shows a large parallax, and if not, the search can be extended to fainter red dwarfs with relative ease, since the search program can deal with many tens of thousands of individual stars.

One important deficiency exists in this, or in almost any, single-site observing program: You can't see the entire sky from the hills east of Berkeley, California. During the course of a year, all the constellations north of the celestial equator (the points directly above Earth's equator) will rise and set, as will the constellations only modestly to the south of the celestial equator, but the regions of the sky close to the south celestial pole (the point directly above Earth's south pole) will never rise at all. To observe all these regions of the sky, you must also observe from a site close to, or south of, the equator. Several modern observatories now exist in Chile and Australia to study the southern skies, but none as yet has an automated program to search for parallax shifts.

If the Leuschner Observatory search fails to reveal Nemesis, the Nemesis theorists will therefore have a simple rejoinder—that Nemesis may well lie in the southernmost third of the skies above us, in a direction outside the Leuschner Observatory's field of view, or may have so low a luminosity that the catalog of the five thousand brightest red-dwarf stars does not include it. A somewhat greater effort can overcome both of these problems, so any such relief will be temporary. If Nemesis fails to emerge from a full-sky survey of parallax shifts, carried out for the twenty thousand red-dwarf stars of greatest apparent brightness, then the Nemesis hypotheses will suffer a downturn, even though Nemesis could exist with sufficiently small mass for its low luminosity to keep it out of this list of brightest red-dwarf stars. The Leuschner Observatory search should be complete at this level by 1987, and plans are being made to have a southern-sky search in operation by that time. The possibility of a full-sky rejection of the Nemesis hypothesis, at least among the five thousand red-dwarf stars of greatest apparent brightness, therefore lies only a few years in the future.

On the other hand, the search may reveal Nemesis! If this is so—if astronomers find the star with the largest parallax shift—reputations will be made, bets paid and pocketed, toasts made and accepted, and textbooks given new printings. Before we contemplate this rush of possibility, let us cast a glance at the other potential method to find Nemesis.

Proper Motions of Stars

As the sun and its neighboring stars orbit around the center of the Milky Way galaxy, almost all the stars move in the same direction, in the same plane, and in nearly circular orbits (plus the additional small up-and-down oscillation that underlies the galactic-carousel hypothesis). However, each star's orbit differs by just a bit from those of its neighbors. Because of these differences, stars do not stay completely abreast of each other, moving in lockstep; instead, although they do so to a first approximation, they slowly approach one another, or diverge, with random motions as seen from any particular star, such as the sun. Through accurate measurements of the stars' positions year after year, we can find the individual motions of the closest stars with respect to the sun, once we allow for the parallax effect. Astronomers still call these individual motions *proper motions*, using the original sense of "proper" (one's own).

The proper motion that we measure for a particular star, typically in seconds of arc (each equal to $1/3,600$ of a degree) per year, will depend on two effects. First, stars with larger velocities with respect to the sun tend to have larger proper motions. Second, for a given velocity with respect to the sun, closer stars will show larger proper motions than more distant stars, because the same motion appears to us as a smaller angle as the stars' distances from us increase. Like the parallax shift, the proper motion of any star with a given velocity with respect to the sun will decrease as the star's distance increases. The star with the largest proper motion, Barnard's Star, ranks fifth in proximity (among *known* stars); it has a fairly large velocity as seen from the sun and is one of the closest stars. The Alpha Centauri stars occupy twelfth place on the list of proper motions; their velocity relative to the sun falls well below that of Barnard's Star, so even though they are the closest known stars, Barnard's Star's much larger velocity with respect to the sun gives it a proper motion almost three times larger than that of the Alpha Centauri system.

As the closest star of all to the sun, Nemesis would have a fine leg up on its proper-motion competitors, but it would have a great handicap as well. The other stars are moving on their own, and have large velocities across our line of sight. Nemesis, however, would be moving in an orbit *around* the sun (more precisely, around the solar system's center of mass). As Nemesis came closer to the sun, it would move more rapidly, as a response to the sun's increased gravitational force. At such times, Nemesis's proximity to the sun and Earth, plus its rapid motion in orbit, would give Nemesis a large proper motion.

On the other hand, when Nemesis was near the farthest

points of its orbit—the current situation according to the hypothesis—it would move only slowly. Calculations based on Newton's laws of gravitation and motion show that Nemesis should now have an orbital speed less than a hundred meters per second. Compare this with the Earth's orbital velocity of 30 kilometers per second and you find that Nemesis's small velocity with respect to the sun, coupled with its proposed distance of more than 3 light years, give Nemesis a predicted proper motion of less than one-thousandth that of Barnard's Star, too small to be measurable by current techniques. The proper-motion search for Nemesis, predicted to fail, might yet be a good idea, since Nemesis *might* exist and could occupy one of the closer points of its orbit. However, in this case the parallax effect would be tremendously enhanced, since the amount of both the parallax shift and the proper motion increase as the distance decreases. In order to find the sun's closest neighboring stars, not yet recognized as such, we ought therefore to invest our effort in parallax-shift rather than in proper-motion studies as we assemble computerized, accurate records of the positions of red-dwarf stars on the sky.

For now, we may simply admire the efforts made to find Nemesis, and reflect that these searches have value whether or not they find a solar companion. Not only will they deal with an important aspect of the theories meant to explain mass extinctions, but they will also significantly improve our ability to measure and to analyze stellar positions. This in turn will give us better knowledge of the sun's closer neighbors, which are mostly red-dwarf stars of low luminosity, hard to find and not considered exciting objects of interest to most astronomers. But as we shall see, such stars may be the single best hope to find other systems with life around them. For those interested in the possibility of finding life outside the solar system, any program that gathers more data about the low-luminosity, low-interest red-dwarf stars has merit, no matter what the final judgment of the Nemesis theory.

5

IMPLICATIONS OF THE SHIVA THEORY

OUR EXAMINATION OF the Shiva theory has led to no hard-and-fast conclusions about either its overall validity or the validity of its four chief components. We can, however, allow ourselves the final speculation of asking what it means to us if the Shiva theory proves correct, in whole or in part.

If a 10-kilometer asteroid or comet caused the extinction of the dinosaurs, this would explain the mystery of how animals that were able to survive for so long suddenly vanished from the earth, and would warn us of the dangers associated with impacts from asteroids and comets. Such impacts are likely to occur every few hundred million years even without any provocation from nearby stars or interstellar clouds; we "need" the Nemesis or galactic-oscillation theories only if we demand that impacts occur every 25 to 35 million years, and on a regular cycle. If the extinction of the dinosaurs proves to be, as seems likely, just one of many mass extinctions during the past 250 million years, then already we must reassess any view of evolutionary biology that does not include these times of tremendous removal of species from Earth.

Implications of Mass Extinctions

Mass extinctions that occur every few tens of millions of years, removing a large fraction of the families of living organ-

147

isms and an even larger fraction of living species, must have played a key role in the evolution of life on Earth. Such extinctions make way for a sudden flowering of new species, a clearing of the undergrowth of organisms that opens immediate opportunities to new forms of life.

Biologists disagree on the precise mechanisms that create new species of organisms, but majority opinion tends toward the view that a region isolated from successful species already in existence is required. Natural selection describes the situation in which individuals of a given species always have some variation and consequently produce different numbers of surviving offspring in response to a particular environmental pressure. Those survival-helping characteristics that can be passed on by inheritance will appear in the offspring, who in turn should do better at producing surviving offspring of their own. As a result, the characteristics of all individuals of a given species can change slowly with time, in response to pressures induced by the environment. However, such gradual evolution cannot increase the total *number* of species significantly.

Natural selection can therefore explain, for example, why giraffes evolved progressively longer necks, in response to the greater reproductive success of those giraffes that could reach higher leaves. A debate exists as to whether such gradual processes of natural selection can explain how entirely new species come into existence, or whether new species appear only as specific, "punctuated" events, in response to sudden changes in the environment of a particular, isolated group of individuals. For this, we probably require some mechanism that isolates relatively small populations of organisms, giving a "new idea" expressed in the small population's genetic variation a chance to flourish. In such isolation, previously unimportant components of the variation within a species may assume new importance, and a new species can appear within a short time, perhaps within a few dozen generations. In a larger group, such components would be swamped by the competition from great numbers of other individuals. This evolutionary model can explain why new species can appear within less than a million years, during which natural selection operates with increased effect on a relatively small population. The newly-evolved species can then stabilize, showing little change for tens of millions, or even hundreds of millions, of years. This model presents a coherent way to deal with the results of mass extinctions, which open tremendous numbers of new opportunities to the survivors. The fossil record shows that each episode of mass extinction has been followed by the rapid appearance of a great number of new species. We may conclude that even if "gradualistic" evolution plays some role in creating new species, "punctuated" bursts of rapid evolution *certainly* occur and may be the only significant way for large numbers of new species to arise.

Modern evolutionary theory does not assume, however, that organisms have developed, or could develop, a response to the *recurrence* of these mass extinctions. No organism can adapt itself through natural selection to be able to survive cataclysmic events that recur on time scales as long as ten million years. We might use the term *species selection* (rather than *natural selection*, which acts among individuals of a particular species) to describe the effect that a nuclear winter, or a comet impact, would have on species. The result would definitely be a discrimination among *species*, but not a continuing pressure that discriminates more successful from less successful *individuals*. In this sense, mass extinctions form part of the evolution of life on Earth. Cockroaches have lasted for 200 million years because they could not only fare well in "ordinary" circumstances but could also survive the periods (if they occurred) of extreme stress that arose at widely spaced intervals.

We intelligent humans, of course, represent the exception to the rule: We *can* look into the future and adapt ourselves to the next mass extinction, if we have the will to do so. If we accept the periodicity of mass extinctions, we face a new time of stressful existence somewhere between 13 and 27 million years from now. For the present, this amounts to saying that *we* have no problem; at some future time, our more cohesive and forward-looking descendants may be sociologically and technologically equipped to confront the challenge of the next impact events. We may even conceive that wrestling with the dilemma of nuclear winter may—if we avoid it—teach us something about how to deal with the effect of cometary impacts.

Implications of Theories to Explain Mass Extinctions

So long as mass extinctions do occur on Earth at intervals of tens of millions of years, it makes little difference whether they are truly periodic in their occurrence, whether they arise from asteroids or comets, or whether random perturbations of aster-oidal orbits, a solar companion, or the sun's galactic oscillation cause impacts to occur. In any event, the existence of mass extinctions must be taken into account in analyzing the history of life on Earth. But our work begins in earnest when we attempt to apply conclusions from the Shiva theory to life that may exist beyond our solar system.

Other planetary systems, we now believe, should have formed around many other stars in our Milky Way galaxy and beyond. As in our own system, cometary planetesimals would have accreted, and then formed agglomerations that became planets and satellites, orbiting a central star. Leftover debris would provide any other system with its own cometary Oort cloud of comets. The system would also have asteroids, solid objects orbiting relatively close to the star, within the region from which volatile compounds have escaped. Passing stars and close encounters with molecular clouds would occasionally divert comets inward, toward the planetary regions, and some asteroids would be diverted into planet-crossing orbits by the gravitational forces from the larger planets in the system.

All such planetary systems should therefore have the potential for occasional asteroid impacts, and since passing stars would perturb comets at about the same frequency throughout the galaxy (somewhat more often in regions closer to the center, where stars crowd more densely, somewhat less often near the galaxy's outer fringes), comets, too, would sometimes make impact craters on those planets with solid surfaces. We can thus expect that *some* mass extinctions occur on any life-bearing planet with a solid crust and an atmosphere, as debris from the impact produces darkness and cold temperatures until settling to the surface, on roughly the same time scale that we calculate for Earth. Gaseous planets, by contrast, may well have developed life forms that float in the atmosphere that constitutes the bulk of the planet. Such forms of life will be basically unaffected by an incoming asteroid or comet, which would simply sink through the planet's thickening veil, perhaps to encounter a solid object wrapped within thousands of kilometers of gas, far below the regions that teem with life.

As a rough and ready rule, we can state that solid planets can have mass extinctions but gaseous planets can not—at least as the result of impacts from preplanetary debris. The galactic-oscillation theory predicts that all solid planets should have mass extinctions that recur at regular intervals. In contrast, the Nemesis theory requires a special situation—a low-mass companion star—to explain the periodic recurrence of mass extinctions. Hence this theory predicts periodic mass extinctions for the solid planets in only a few planetary systems. The Nemesis theory thus forces us to adopt a conception of the evolution of life on Earth according to which our history would be special, not representative of what has occurred on similar planets orbiting stars similar to ours.

Nemesis Versus the Galactic-Oscillation Theory

The Nemesis theory singles out our solar system and our history of life from the general run of affairs in the Milky Way, making us not unique—for surely *some* stars have low-mass companions in immense, elongated orbits—but highly unusual. Skeptics might say that the largest single drawback to the Nemesis theory is its requirement that our sun and planets appear to have a nonrepresentative role in the cosmos. Astronomy has made great strides by abandoning the pre-Copernican notion that the universe centers on our planet, and by testing the hypothesis that our planet is an average one, in orbit around an average star, in a representative galaxy, in a representative position within the universe. To abandon this notion, even in part, demands strong evidence.

If the evidence appears—if astronomers discover Nemesis—then scientists will be ready to shift their views. We could, after all, be special, just as our instincts insist. But if so, then a clear implication arises: To the extent that the periodic mass extinctions caused by Nemesis have influenced evolution on Earth, we cannot expect to find a similar history in the vicinity of most stars with planets. Only where other "Nemeses" have likewise cleared the way with a recurring cycle of destruction and renewal should we anticipate anything similar to our own story. If this pattern proves necessary for complex forms to arise within a few billion years—if other planets have not yet progressed beyond the equivalent of single-celled plants and animals, because evolutionary openings like those from our mass extinctions have yet to emerge—then we may be effectively alone in the galaxy, simply because of the rarity of situations like our own, assuming that we *do* have a companion star that triggers mass extinctions.

The UFO Debate: Where Is Everybody?

This situation would provide a reasonable answer to the question that you may think doesn't puzzle scientists: Where is everybody? This question, first posed by the great physicist Enrico Fermi, refers to the conclusion that since our galaxy (not to mention trillions upon trillions of other galaxies) contains hundreds of billions of stars, each roughly like our sun, and each quite likely to have planets in orbit around it, if life arises from a mixture of the elements typical of these stars and their planets, within a range of temperature conditions only modestly restrictive, then we should expect to find our galaxy teeming with life. Even if only one star in a thousand had a planet on which life could arise, we would expect to find hundreds of millions of planetary systems with life. And if these planets do produce "intelligent" life, given "world enough and time," then intelligent life, defined as creatures capable of communicating over interstellar distances, should be rampant in the Milky Way. On *some* of these planets, civilizations with both the means and the desire to pay for interstellar space flight should have arisen, and some of these should visit Earth, perhaps not every day or every year, but often enough that we should have proof that they exist.

Despite the claims of sensationalist newspapers, such evidence is conspicuously absent from the record. Reports of UFO sightings have turned out to furnish either extremely strange tales by unreliable observers or only moderately strange tales from completely reliable sources. Since any extraterrestrial visitors could easily obtain tremendous, detailed publicity (national television, for example), those who consider UFOs to be extraterrestrial spacecraft are led to propose an extraterrestrial "shyness," an unwillingness to have a good, long talk, entirely out of keeping (so far as we can speculate about the psychology of extraterrestrials) with the drive to explore the Milky Way. The "shyness" theory has also been modified to the theory that Earth is the zoo of more advanced forms of life, a good explanation of politicians' behavior but not a likely one without better evidence. (Note, however, that this theory includes the prediction that such evidence will be hard to acquire!)

Most scientists would resolve the "Fermi paradox" (Where is everybody if life is so prevalent?) either by stressing the enormous energy needed for interstellar space flight, so that many civilizations could exist without many close encounters, or by concluding that "intelligent" life must be less abundant than a straightforward extrapolation from Earth's history would imply. The Nemesis theory falls in the second category, and it carries a harsh message to those who hope to search for extraterrestrial life: Our solar system has special characteristics, good for us but absent in most planetary systems, so don't expect an easy job in the attempt to communicate with other forms of life.

The other variant of the Shiva theory tells us just the reverse: All stars oscillate through the galactic plane as they circle the galaxy, so all planetary systems should have periodic impacts and mass extinctions. We are indeed representative of the way in which life has evolved and should expect to find civilizations something like ours throughout the Milky Way. It only remains to explain why none are visiting us, and how we can nonetheless establish contact through radio and television messages. This preserves the conventional view of our solar system as average rather than nearly unique.

The Nemesis theory, and its competitors that seek to explain mass extinctions, thus prove to have tremendous implications in our search for life in the cosmos. We must merely discover which implications are correct by resolving which, if any, of the explanations for periodic mass extinctions correspond to reality, and which represent glorious excursions into an imaginary universe. The scientists who have laid these theories before us, since they *are* scientists, have also shown us the ways in which we can discriminate between more correct and less correct theories. Tread softly for you tread on their hypotheses.

Envoi: What Makes a Theory "Scientific"?

The reader who has come this far through our tale of mass extinctions, periodicity, cometary showers, galactic oscillations, and the solar-companion star no longer qualifes as a perfect representative of the public at large when it comes to scientific interests. Clearly the selection process implied in reading about a scientific hypothesis in detail has left behind, among others, those who strongly suspect, or firmly believe, that scientists often take our money and spend it on pointless cerebration, or on more wasteful space hardware. Such people might have a field day with the Shiva theory; here we find fossil experts debating whether the extinctions of tens of millions of years ago came in cycles; whether such extinctions might have arisen from impacts or from more gradual environmental changes; whether the sun's comets number in the tens of trillions or only in the trillions; whether the comets might or might not be perturbed in sufficiently large numbers to have had significant effects on Earth, millions of years ago; and whether such perturbations might be caused by a previously unknown star that, if detected, has no role to play for another 13 million years. For this we should pay good salaries, not to mention funding sizable observational programs?

But even the most aroused scientific opponent of the Shiva theory—one who finds it ridiculous that his colleagues spend their energy inventing new stars to explain unrecorded cometary showers as the cause of unverified periodic behavior in the largely incomplete fossil record—knows that scientific progress has often occurred through just such far-ranging, apparently wild-eyed speculation. And just because far more often, little if any progress results from such activity does not mean that its importance is diminished; you cannot make bold advances without bold failures. Most important of all is that the Shiva theorists are "doing science"—creating hypotheses based on a limited set of data that can be tested for survival as new data, which the theory itself suggests ought to be gathered, are obtained. If the Shiva theorists insisted that they *must* be right, that no experiment or data could disprove the logic and grandeur of their hypothesis, then we would be wasting money to support them. Such an attitude would be reminiscent of the late great pseudoscientist Immanuel Velikovsky, who developed a complex hypothesis, in some ways reminiscent of the Shiva theories, that Earth had undergone near-collisions with the planets Mars and Venus, that these encounters could explain many of the tales in the Bible and in other ancient documents, and that our view of the solar system must be largely revised to take account of his "worlds in collision," planets wandering from what we take to be stable orbits.

Velikovsky was not a scientist, for the simple reason that he was "always right." Indeed, his world view did not allow for the possibility of error on his part: He could not, and did not, state the facts that would disprove his hypotheses if discovered. From a conventional astronomical view, Velikovsky was dead wrong when he first published his theory, in 1950. Velikovsky called for planets—not just comets or asteroids—to have wandered tens of millions of kilometers from their present orbits within the past few thousand years, and for Mars and Venus to have been expelled from Jupiter shortly before that. By 1950 we already knew that astronomical records from Mesopotamia showed that Venus must have had its present orbit, or one close to it, before the time that Velikovsky demanded that it nearly collided with Earth, and our knowledge of the dynamics of how planets move, and of their composition, ruled out the dancing motions that Velikovsky demanded.

Although the case against Velikovsky's theories grew far stronger as we sent spacecraft to explore the solar system, landing on Mars and Venus and passing close by Jupiter, this was never really the fundamental resolution of the issue: Velikovsky and his supporters invented new mechanisms to preserve the basic theory. Even this can be acceptable in science, so long as the new inventions are clearly delineated, if the hypothesizer is willing to outline what tests can disprove the theory. Velikovsky never was willing, and he gradually adopted the martyr's role after some scientists foolishly tried to prevent the publication of his first book in 1950. Like Galileo, Velikovsky was scorned by the establishment of his time; like Galileo, too, he knew that his critics were little men, destined to the dustbin of history. Unlike Galileo, Velikovsky could never present his theories in a way that allowed reasonable people to test them. This is in some ways more important than that Galileo's theories proved to be right and Velikovsky's wrong. Galileo was a scientist, though an early and an extremely prickly one; Velikovsky stood outside science, refusing to accept its critical ability to refute some theories at the expense of others.

People who simply demand results may not consider important the distinction between scientists' and pseudoscientists' procedures, but scientists do. By constructing an environment—the world of science—in which theories survive not because they are emotionally satisfying but because they fit into the existing framework more successfully than competing theories, scientists have created the potential for anyone within their purview to make important advances in our collective knowledge. Individuals make the theories; the social structure of science does the testing. If you don't accept this principle, you don't belong to the scientific community. This does not mean that your ideas must be wrong, only that you (and they) won't be taken seriously. No one guarantees that your ideas will always receive serious consideration in any case, but there is no hope if you don't "think like a scientist"—accept the proposition that your ideas may or may not be right, and look for ways to prove the former and reject the latter.

The preceding discussion cannot be taken to prove that all theorizing within the scientific framework is a good thing; no doubt some of it is a waste of time, and might be clearly identified as such to almost everyone's satisfaction. But the theorizing about mass extinctions, vague and unsettled though it may be, is a good example of just what scientists have to do to make progress.

The data on mass extinctions, and on the Cretaceous-Tertiary boundary layer, call for explanation. Maybe these explanations fit together, and maybe they don't, but the question ought to be studied. Extinctions may come in cycles, or they may not, but the desire to assemble more data on extinctions and to analyze it deserves support, not because it will do much for our style of life but because it will help us to understand better the history of life on Earth. Comets may strike the earth with greater likelihood at some periods, and that is a phenomenon worthy of investigation no matter what the provoking cause, because comets are some of the original pieces of the solar system, and we ought to try to discover how the sun and its planets formed. Finally, the debate over whether Nemesis exists, and what other perturbations exist from regions relatively close to the solar system in the Milky Way, will help us to uncover new facts that give us a clearer understanding of our galactic environment. Such progress was never intended to yield results beyond the satisfaction of unraveling some of the mysteries of the cosmos, but it is remarkable how thinking about one part of the universe often gives useful clues to other parts, and some of those parts have practical values on Earth. For example, the impact scenario could yield useful insight into the nuclear-winter effect, a hazard that could make the possibility of future "mass extinctions" irrelevant to human life on Earth.

From some aspects, science needs no justification: We all should rejoice in the increase in the totality of what we know. When it comes to spending more or less money, however, we naturally ask harder questions: Which parts of science deserve more support? How much for science and how much for highways? These are rightly political questions, in the largest sense, with political answers. The theory that mass extinctions recur periodically may have much to say about life on our planet, but it has small justification for spending large sums of money. Fortunately, even the small sums available for everyday research in biology, paleontology, geology, and astronomy may soon settle the major controversies of the past few years, and may in this way point us to the next such theoretical debate, whatever that may be. For scientists, this prospect suffices to inspire more effort, aimed at enlarging what we know and what we can hope to find out.

FURTHER READINGS

The recent advent of the Shiva theory limits further reading on details not covered by this book to journals in which the theory's seminal components were first proposed.

The best introduction to the asteroidal-impact hypothesis for the Cretaceous-Tertiary extinctions is:
"Experimental Evidence That an Asteroid Impact Led to the Extinction of Many Species 65 Million Years Ago," by Luis Alvarez. *Proceedings of the National Academy of Sciences*, January 1983, vol. 80, p. 627.

The claim for periodic behavior in the mass-extinction data appeared in the same journal:
"Periodicity of Extinctions in the Geologic Past," by David M. Raup and J. John Sepkoski, Jr. *Proceedings of the National Academy of Sciences*, February 1984, vol. 81, p. 801.

A fine description of the role that mass extinctions can play in evolution is:
"The Ediacaran Experiment," by Stephen Jay Gould. *Natural History*, February 1984, vol. 93, no. 2, p. 14.

A rebuttal to the notion that dinosaurs died as rapidly as claimed by impact theorists appears in:

"Late Cretaceous Extinctions," by J. David Archibald and William A. Clemens. *American Scientist*, July–August 1982, vol. 70, no. 4, p. 377.

The key paper on the nuclear-winter effect is:
"Nuclear Winter: Global Consequences of Multiple Nuclear Explosions," by R. P. Turco, O. B. Toon, T. P. Ackerman, J. B. Pollack, and Carl Sagan. *Science*, 23 December 1983, vol. 222, p. 1283.

Immediately following this paper is another good one:
"Long-term Biological Consequences of Nuclear War," by Paul R. Ehrlich et al. *Science*, 23 December 1983, vol. 222, p. 1293.

At a more advanced level, the exposition of the galactic-oscillation model appears in:
"Geological Rhythms and Cometary Impacts," by Michael Rampino and Richard Stothers. *Science*, 21 December 1984, vol. 226, p. 1427.

The Nemesis theory is presented, also at a more advanced level, in:
"Are Periodic Mass Extinctions Driven by a Distant Solar Companion?" by Daniel P. Whitmire and Albert A. Jackson IV. *Nature*, 19 April 1984, vol. 308, p. 713.

. . . and in:
"Extinction of Species by Periodic Comet Showers," by Marc Davis, Piet Hut, and Richard A. Muller. *Nature*, 19 April 1984, vol. 308, p. 715.

The latter paper is followed by the Nemesis theorists' analysis of the impact-crater data:

"Evidence From Crater Ages for Periodic Impacts on the Earth," by Walter Alvarez and Richard A. Muller. *Nature*, 19 April 1984, vol. 308, p. 718.

Views critical of the impact explanation of mass extinctions appear in two science-news articles:

"Cratering Theories Bombarded," by Paul Weissman. *Nature*, 7 March 1985, vol. 314, p. 17.

"Periodic Extinctions and Impacts Challenged," by Richard A. Kerr. *Science*, 22 March 1985, vol. 227, p. 1451.

The articles listed above contain many more references in their bibliographies, some much more arcane than others.

INDEX

163